Environmental Health Criteria 44

MIREX

Published under the joint sponsorship of
the United Nations Environment Programme,
the International Labour Organisation,
and the World Health Organization

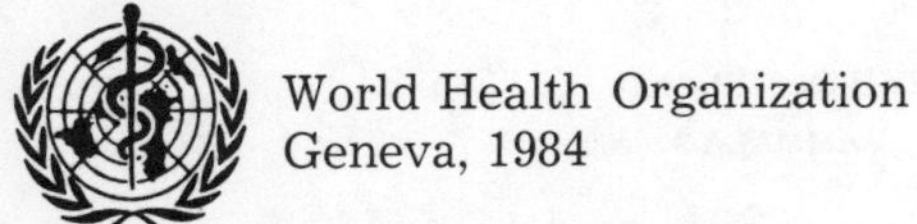
World Health Organization
Geneva, 1984

The **International Programme on Chemical Safety (IPCS)** is a joint venture of the United Nations Environment Programme, the International Labour Organisation, and the World Health Organization. The main objective of the IPCS is to carry out and disseminate evaluations of the effects of chemicals on human health and the quality of the environment. Supporting activities include the development of epidemiological, experimental laboratory, and risk-assessment methods that could produce internationally comparable results, and the development of manpower in the field of toxicology. Other activities carried out by the IPCS include the development of know-how for coping with chemical accidents, coordination of laboratory testing and epidemiological studies, and promotion of research on the mechanisms of the biological action of chemicals.

ISBN 92 4 154184 9

The designations employed and the presentation of the material in this publication do not imply the expression of any opinion whatsoever on the part of the Secretariat of the World Health Organization concerning the legal status of any country, territory, city or area or of its authorities, or concerning the delimitation of its frontiers or boundaries.

The mention of specific companies or of certain manufacturers' products does not imply that they are endorsed or recommended by the World Health Organization in preference to others of a similar nature that are not mentioned. Errors and omissions excepted, the names of proprietary products are distinguished by initial capital letters.

PRINTED IN FINLAND

84/6277 – VAMMALA – 5500

CONTENTS

Page

TASK GROUP MEETING ON ENVIRONMENTAL HEALTH CRITERIA FOR ORGANOCHLORINE PESTICIDES OTHER THAN DDT (CHLORDANE, HEPTACHLOR, MIREX, CHLORDECONE, KELEVAN, CAMPHECHLOR)

Members

Dr Z. Adamis, National Institute of Occupational Health, Budapest, Hungary

Dr D.A. Akintonwa, Department of Biochemistry, Faculty of Medicine, University of Calabar, Calabar, Nigeria[a]

Dr R. Goulding, Chairman of the Scientific Sub-committee, UK Pesticides Safety Precautions Scheme, Ministry of Agriculture, Fisheries & Food, London, England (Chairman)

Dr S.K. Kashyap, National Institute of Occupational Health (Indian Council of Medical Research), Meghaninager, Ahmedabad, India

Dr D.C. Villeneuve, Environmental Contaminants Section, Environmental Health Centre, Tunney's Pasture, Ottawa, Ontario, Canada (Rapporteur)

Dr D. Wassermann, Department of Occupational Health, The Hebrew University, Haddassah Medical School, Jerusalem, Israel (Vice-Chairman)

Representatives of Other Organizations

Dr C.J. Calo, European Chemical Industry Ecology and Toxicology Centre (ECETOC), Brussels, Belgium

Mrs M.Th. van der Venne, Commission of the European Communities, Health and Safety Directorate, Luxembourg

Dr D.M. Whitacre, International Group of National Associations of Agrochemical Manufacturers (GIFAP), Brussels, Belgium

Secretariat

Dr M. Gilbert, International Register for Potentially Toxic Chemicals, United Nations Environment Programme, Geneva, Switzerland

---

a Unable to attend.

Secretariat (contd).

Mrs B. Goelzer, Division of Noncommunicable Diseases, Office of Occupational Health, World Health Organization, Geneva, Switzerland

Dr Y. Hasegawa, Division of Environmental Health, Environmental Hazards and Food Protection, World Health Organization, Geneva, Switzerland

Dr K.W. Jager, Division of Environmental Health, International Programme on Chemical Safety, World Health Organization, Geneva, Switzerland (Secretary)

Mr B. Labarthe, International Register for Potentially Toxic Chemicals, United Nations Environment Programme, Geneva, Switzerland

Dr I.M. Lindquist, International Labour Organisation, Geneva, Switzerland

Dr M. Vandekar, Division of Vector Biology and Control, Pesticides Development and Safe Use Unit, World Health Organization, Geneva, Switzerland

Mr J.D. Wilbourn, Unit of Carcinogen Identification and Evaluation, International Agency for Research on Cancer, Lyons, France

NOTE TO READERS OF THE CRITERIA DOCUMENTS

While every effort has been made to present information in the criteria documents as accurately as possible without unduly delaying their publication, mistakes might have occurred and are likely to occur in the future. In the interest of all users of the environmental health criteria documents, readers are kindly requested to communicate any errors found to the Manager of the International Programme on Chemical Safety, World Health Organization, Geneva, Switzerland, in order that they may be included in corrigenda, which will appear in subsequent volumes.

In addition, experts in any particular field dealt with in the criteria documents are kindly requested to make available to the WHO Secretariat any important published information that may have inadvertently been omitted and which may change the evaluation of health risks from exposure to the environmental agent under examination, so that the information may be considered in the event of updating and re-evaluation of the conclusions contained in the criteria documents.

* * *

A detailed data profile and a legal file can be obtained from the International Register of Potentially Toxic Chemicals, Palais des Nations, 1211 Geneva 10, Switzerland (Telephone no. 988400 - 985850).

ENVIRONMENTAL HEALTH CRITERIA FOR MIREX

Following the recommendations of the United Nations Conference on the Human Environment held in Stockholm in 1972, and in response to a number of World Health Assembly Resolutions (WHA23.60, WHA24.47, WHA25.58, WHA26.68), and the recommendation of the Governing Council of the United Nations Environment Programme, (UNEP/GC/10, 3 July 1973), a programme on the integrated assessment of the health effects of environmental pollution was initiated in 1973. The programme, known as the WHO Environmental Health Criteria Programme, has been implemented with the support of the Environment Fund of the United Nations Environment Programme. In 1980, the Environmental Health Criteria Programme was incorporated into the International Programme on Chemical Safety (IPCS). The result of the Environmental Health Criteria Programme is a series of criteria documents.

A WHO Task Group on Environmental Health Criteria for Organochlorine pesticides other than DDT met in Geneva from 28 November to 2 December 1983. Dr K.W. Jager opened the meeting on behalf of the Director-General. The Task Group reviewed and revised the draft criteria document and made an evaluation of the health risks of exposure to mirex.

The drafts of this document were prepared by Dr D.C. Villeneuve of Canada and Dr S. Dobson of the United Kingdom.

The efforts of all who helped in the preparation and finalization of the document are gratefully acknowledged.

* * *

Partial financial support for the publication of this criteria document was kindly provided by the United States Department of Health and Human Services, through a contract from the National Institute of Environmental Health Sciences, Research Triangle Park, North Carolina, USA - a WHO Collaborating Centre for Environmental Health Effects.

1. SUMMARY AND RECOMMENDATIONS

1.1 Summary

1.1.1 Identity, properties and analytical methods

Mirex ($C_{10}Cl_{12}$) is a white crystalline odourless solid; it is an extremely stable substance.

Gas chromatography with electron capture detection is the analytical method most commonly used for its determination.

1.1.2 Uses and sources of exposure

Mirex is mainly used as a flame-retardant and as a stomach insecticide, mainly formulated into baits, for the control of ants, especially fire ants and harvester ants. The USA appears to be the main country in which mirex was used for pest control, but this use was discontinued in 1978.

The same chemical substance is also used, under the name Dechlorane, as a fire retardant in plastics, rubbers, paints, etc. This application is not restricted to the USA.

A known source of exposure for the general population is food. However intake from this source is below the promulgated tolerance levels.

No data are available on occupational exposures to mirex.

1.1.3 Environmental concentrations and exposures

Mirex is one of the most stable chemicals in use today. Biodegradation by microorganisms does not take place except, occasionally, under anaerobic conditions, and, even then, at a slow rate.

Photodegradation under the influence of UV radiation is slow, photomirex (8-monohydromirex) being the major degradation product. The environmental half-life of mirex is of the order of many years, and its breakdown products are equally stable.

Because it is practically insoluble in water, sediments act as a sink for mirex that enters waterways.

Mirex bioaccumulates at all trophic levels and is biomagnified through food chains.

Long-term toxicity, with delayed onset of toxic effects and mortality is uniformly high. Mirex is toxic for a range of aquatic organisms, with crustacea being particularly sensitive.

Thus, it appears that mirex presents a long-term environmental hazard.

1.1.4 Kinetics and metabolism

Following oral ingestion, mirex is only partly absorbed into the body and the remainder - depending on the dose administered - is excreted unchanged in the faeces. Mirex can also be absorbed following inhalation and via the skin.

It is a lipophilic compound and, as such, is stored in adipose tissue to a greater extent than in any other tissue. Mirex is transferred across the placenta to the fetus and is excreted with the milk.

Mirex does not appear to be metabolized to any extent in any animal species investigated. Its elimination from the body is slow. Depending on the species tested, its half-life in the body is several months.

1.1.5 Studies on experimental animals

Mirex is moderately toxic in single exposures. In long-term studies, far lower daily dosages (1 mg/kg diet) have led to liver hypertrophy with morphological changes in the liver cells, and induction of mixed-function oxidases.

It is fetotoxic and teratogenic.

Mirex is not generally active in short-term tests for genetic activity.

Mirex is carcinogenic for both mice and rats.

1.1.6 Effects on man

No reports on accidental poisoning or occupational exposure and occupational health effects are available.

1.2 Recommendations

1. Surveillance should be maintained over any future production, transport, and disposal of mirex and the nature and extent of both its agricultural and non-agricultural use.

2. Levels of mirex in the environment should continue to be comprehensively monitored.

2. IDENTITY, PHYSICAL AND CHEMICAL PROPERTIES, AND ANALYTICAL METHODS

2.1 Identity

Chemical structure:

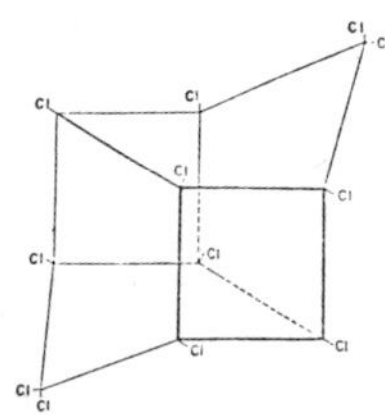

Molecular formula:	$C_{10}Cl_{12}$
CAS chemical name:	1,1a,2,2,3,3a,4,5,5,5a,5b,6-dodeca-chloroocta-hydro-1,3,4-metheno-1H-cyclobuta[cd]pentalene
Synonyms:	dodecachloropentacyclo[5.2.1.$0^{2\,6}0^{3\,9}$ $0^{5\,8}$]decanedodecachlorooctahydro-1,3,4-metheno-2H-cyclo-buta[cd] pentalene
Trade names:	Dechlorane, Ferriamicide, GC 1283
CAS registry number:	2385-85-5
Relative molecular mass:	545.5

2.2 Physical and Chemical Properties

Mirex is a white crystalline, odourless solid with a melting point of 485 °C. It is soluble in several organic solvents including tetrahydrofuran (30%), carbon disulfide (18%), chloroform (17%), and benzene (12%), but is practically insoluble in water (US NRC, 1978). It has a vapour pressure at 25 °C of $3 \cdot 10^{-7}$ mm (IARC, 1979). Vapour pressures at other temperatures can be found in Matsumura (1975).

Mirex is considered to be extremely stable (US NRC, 1978). It does not react with sulfuric, nitric, hydrochloric or other common acids and is unreactive with bases, chlorine or ozone. Despite its stability, reductive dechlorination of mirex can be brought about by reaction with reduced iron porphyrin or more effectively by vitamin B_{12} (Schrauzer & Katz, 1978). Slow partial decomposition will also result from ultraviolet (UV) irradiation in hydrocarbon solvents or from gamma rays (Lane, 1973; Baker & Applegate, 1974). Dechlorination by UV irradiation yields photomirex (8-monohydromirex) as a major product (Alley et al., 1974; Mehendale, 1977a) and this may represent the fate of most of the mirex in the environment (Mirex Advisory Committee, 1972[a]; Carlson et al., 1976).

Mirex is quite resistant to pyrolysis; decomposition begins at 525 °C (Kennedy et al., 1977), and 99 - 98% combustion is accomplished at 700 °C within 1 second (Wilkinson et al., 1978). Hexachlorobenzene is a major pyrolysis product with lesser amounts of carbon monoxide, carbon dioxide, hydrogen chloride, chlorine, carbon tetrachloride, and phosgene given off as vapour.

According to US NRC (1978), technical grade preparations of mirex contain 95.19% mirex and 2.58% chlordecone; the rest of the composition was not specified. The term "mirex" is also used to refer to a bait comprising corncob grits, soya bean oil, and mirex (IARC, 1979). Insect bait formulations for aerial application containing 0.3 - 0.5% mirex and fire ant formulations containing 0.075 - 0.3% mirex have also been used in the USA (IARC, 1979).

2.3 Analytical Methods

Several analytical procedures, used for the determination of mirex, are summarized in Table 1. Other methods used include gel-permeation chromatography, and gas chromatography using an electrolytic conductivity detector (IARC, 1979). Mirex can be analysed in the presence of PCBs by using nitration procedures (Task Force on Mirex, 1977), perchlorination (Hallett et al., 1976), or photodegradation (Lewis et al., 1976). High pressure liquid chromatographic (HPLC) methods have also been used for the separation and quantitation of mirex and PCBs (Task Force on Mirex, 1977).

---

[a] Report to US Environmental Protection Agency.

Table 1. Methods for the determination of mirex

Sample type or medium	Sampling method, Extraction/cleanup[a]	Analytical method[a]	Limit of detection	Reference
general		GLC	0.2 pg	Hartman (1971)
air	trap on polyurethane foam, extract with hexane-ether, wash	GC/ECD	0.1 ng/m^3	Lewis et al. (1977)
water				
rural potable	extract with hexane, CC	GC/ECD	0.01 µg/litre	Sandhu et al. (1978)
fresh & salt		GC/ECD	0.001 µg/litre	Markin et al. (1974b)
soil & sediment	extract with petroleum ether or acetone-petroleum ether, CC	GC/ECD	3-6 µg/kg	Bevenue et al. (1975)
fruit & vegetables	extract with acetonitrile or aq. acetonitrile, liquid/liquid partition, CC	GC/ECD, TLC, PC	-	Horwitz (1975)
fatty products & fish	mix with florisil, extract with acetonitrile, liquid/liquid partition, CC	GC/ECD	-	Bong (1975, 1977)

Table 1 (contd).

Sample type or medium	Sampling method, Extraction/cleanup[a]	Analytical method[a]	Limit of detection	Reference
catfish	grind with anhyd. sodium sulfate, extract with hexane, CC	GC/ECD	10 µg/kg	Collins et al. (1973)
biological material, wildlife	grind with anhyd. sodium sulfate, extract with hexane-isopropanol, wash with water, CC	GC/ECD	1 µg/kg	Collins et al. (1974)
wildlife	mix with florisil, extract with 5% water in acetonitrile, liquid/liquid extraction, CC, chlorinate	GC/FID/ECD/CD/GC/MS	-	Hallett et al. (1976)

[a] GLC - gas-liquid chromatography.
GC/ECD- gas chromatography/electron capture detection.
CC- column chromatography.
TLC- thin layer chromatography.
PC- paper chromatography.
FID- flame ionization detection.
CD- conductivity detection.
MS- mass spectrometry.

3. PRODUCTION, USES, TRANSPORT AND DISTRIBUTION

3.1 Production and Uses

Mirex was first synthesized in 1946 by Prins but was not used in pesticide formulations until 1955.

Mirex is made by the dimerization of hexachlorocyclopentadiene in the presence of aluminum chloride (IARC, 1979). It is a stomach insecticide with little contact activity. The insecticidal use of mirex has been largely focused on the control of the imported fire ant *Solenopsis saevissima richteri*, in southeastern USA. The imported fire ant was introduced into the USA at the beginning of this century and for the first twenty years confined itself to the area around the port of Mobile, Alabama. However, a second wave of a closely related species (*Solenopsis invicta*) appeared in the late 1920s and spread throughout the south of the USA. Since then, the imported fire ant has infested some 76 million hectares in the southern USA (Gunby & Preston, 1979). This pest can pose a nuisance as it can deliver a severe sting which often results in secondary infection. In addition, the mounds produced by the ants make farming difficult and can cause damage to farm machinery. To combat the problem, approximately 250 000 kg of mirex was applied to fields during 1962-75 (US NRC, 1978). Most of the mirex was in the form of 4X mirex bait, which consists of 0.3% mirex in 14.7% soybean oil mixed with 85% corncob grits. Preparations also came in 2X and 1X baits, which contained 0.15 and 0.1% mirex, respectively. Application of the 4X bait was designed to give a coverage of 4.2 g mirex/ha and was delivered by aircraft, helicopter or tractor. Another form of bait consists of microencapsulated mirex in soybean oil (Markin et al., 1975). Normal application rates are 750 mg active ingredient/kg for fire ant baits and 1500 mg/kg for harvester ant baits.

Other pests are also sensitive to mirex, including the western hamster ant, the yellow-jacket and the Texas leaf-cutting ant (Mirex Advisory Committee, 1972).[a] Mirex bait has been applied to pineapple-growing areas in Hawaii to control mealy bug, under permit from US EPA since 1970 (Bevenue et al., 1975).

In 1971, the US EPA cancelled all federal regulations permitting the use of mirex pending release of an environmental impact study. New regulations, issued by the US EPA in 1972, authorized the restricted use of mirex by permit

---

[a] Report to US Environmental Protection Agency.

only. Allied Chemical Corporation, which at the time was the sole producer of bait formulations, sold the registration for mirex and the right to produce, for one dollar, to the Mississippi Department of Agriculture (Pesticide Chemical News, 1976). During the same year, the US EPA ordered a phasing-out of the use of mirex for pest control and brought in a ban with exemptions on June 30, 1978. Mississippi has since been trying to gain approval for a new compound under the generic name Ferriamicide. This bait contains long-chain alkyl amines and ferrous chloride in addition to mirex. With this composition, 80 - 90% of the mirex was claimed to degrade within 30 days, compared with the normal break-down time of 5 - 10 years (Kaiser, 1978). However, a Canadian study conducted in 1979 (Villeneuve et al., 1979a) demonstrated that photomirex, a major break-down product of Ferriamicide, was considerably more toxic than mirex itself. Hence, the US EPA has withheld permission for the Mississippi Pest Control Program, pending review of the Canadian study (Gunby & Preston, 1979). The literature indicates that the USA may be the only country to have used mirex in pest control.

Mirex, under the name Dechlorane, is also used as a fire retardant in plastics, rubber, paint, paper, and electrical goods, and as a smoke-generating compound, when combined with zinc oxide and powered aluminum. Statistics show that between 1959 and 1975, 400 000 kg of mirex and 1 500 000 kg of Dechlorane were sold, of which 74% was used in the USA for non-agricultural purposes (US NRC, 1978). Recently, non-agricultural mirex has been replaced in part by compounds such as Dechlorane plus, Dechlorane 4070, 510, 602, 603, and 604, all of which have similiar fire retardant properties. No recent consumption data for mirex in non-agricultural applications could be obtained.

Unfortunately, complete information on the quantities of mirex produced in the USA and its fate is not available. In fact, as much as half of the mirex used between 1962-73 cannot be accounted for (US NRC, 1978). Little information is also available on world-wide production and use, but patents for the use of mirex exist in several countries including Belgium, France, the Federal Republic of Germany, Japan, the Netherlands, and the United Kingdom (Task Force on Mirex, 1977).

3.2 Transport and Distribution

(a) Air

There is no documentation concerning air-borne mirex contamination in the literature. It is reasonable to assume, however, that facilities involved in the production of mirex and its by-products may have released significant levels of mirex dust into the atmosphere within and immediately surrounding the plants. The Task Force on Mirex (1977) suggested that aerial transport could possibly be involved in the contamination of non-target organisms in untreated areas.

(b) Water

As mentioned earlier, mirex has a very low solubility in water and if concentrations exceed 1 μg/litre, mirex would be associated with the particulate matter in the water rather than with the water itself. It has been demonstrated that mirex can be translocated to water bodies from adjacent agricultural land (Borthwick et al., 1973; Spence & Markin, 1974; Tagatz et al., 1975).

(c) Soil

When mirex is used in pesticide formulations, it is generally in the form of a bait and, thus, not applied directly to the soil. After application of 0.04 g mirex bait/ha, mirex residues in the soil ranged from 0.1 - 10 μg/kg (Mirex Advisory Committee, 1972).[a] Jones & Hodges (1974), found that only 6.6% of the mirex from bait leached into the top 1.5 cm of the soil in a test plot after 6 months exposure to sun and rain. This is supported by field studies such as the Residue Monitoring Program on Hawaii (Bevenue et al., 1975).

Sediments can act as sinks for the small amount of mirex that is leached and deposited via run-off. Residue levels typically mimic soil levels, and are normally quite low. However, in Lake Ontario, levels of mirex as high as 40 μg/kg have been reported in sediments near the Oswego and Niagara areas. These high levels have been attributed to the dumping of mirex in the rivers and not to soil run-off or leaching (Task Force on Mirex, 1977).

Very little information is available concerning the leaching of Dechlorane from landfill sites or disposal of flame retardant material, but this may also represent an important source of contamination.

---

[a] Report to US Environmental Protection Agency.

3.3 Abiotic degradation

Mirex is considered to be one of the most stable pesticides in use today (Baker & Applegate, 1974). Several conditions under which reductive dechlorination of mirex will occur have been given in section 2 of this report. The most significant factors involved in abiotic degradation in the environment are ultraviolet light and gamma irradiation.

Conversion of mirex to the 8-monohydro derivative (photomirex) was shown to occur when mirex was exposed to sunlight (Gibson et al., 1972). Mirex has also been shown to undergo photolytic dechlorination in some organic solvents (Dilling & Dilling, 1967; Alley et al., 1973) and mallard duck eggs (Lane et al., 1976) when exposed to UV radiation. The primary photodegradation product in these cases was photomirex with lesser amounts of 5,8 dihydro-mirex. Carlson et al. (1976) showed that from 16 to 19.5% of the total mirex-related residues from soil samples, recovered 12 years after treatment at 1.12 kg/ha, was photomirex. Lesser amounts of chlordecone (3.1 - 6.3%), 10-monohydro-mirex, and 2 isomers of dihydro-mirex were also present. When 4X mirex bait was exposed to intense UV radiation for 19.5 h, similar degradation patterns were found, the major degradation product being photomirex (19.9%), with lesser amounts of chlordecone (0.2%) and other derivatives (Carlson et al., 1976).

The half-life of mirex dispersed in water under intense UV radiation at 90 - 95 °C was 48.4 h (similar to DDT: 42.1 h); this was rather long compared with that of dieldrin (11.5 h) (Knoevenagel & Himmelreich, 1976).

Several photodegradation products that occur in the environment include 10-monohydromirex, 8-monohydromirex, 5,10-dihydromirex, chlordecone and 2,8-dihydromirex (Alley et al., 1973, 1974; Baker & Applegate, 1974; Ivie et al., 1974a; Carlson et al., 1976). These compounds have been demonstrated to occur in the laboratory and under field conditions as a consequence of irradiation. Levels in twelve-year-old experimental plots in Mississippi suggested that mirex had an environmental half-life of many years (Carlson et al., 1976). As mentioned previously, an environmental half-life of 5 - 10 years has been cited in other studies (Carlson et al., 1976; US NRC, 1978). Data collected from a 5-year-old aircraft crash site in which a cargo of bait was dumped in a shallow pond, produced similar results (Andrade & Wheeler, 1974b). The evidence to date suggests that slow partial photo-degradation is likely to be the ultimate fate of mirex in the environment.

3.4 Biodegradation

Mirex is very resistant to microbiological degradation and is only slowly dechlorinated to a monohydro derivative by anaerobic microbial action in sewage sludge (Andrade et al., 1974a, 1975) and by enteric bacteria in monkeys (Stein et al., 1976). There have been no reports of evidence of metabolic degradation by soil microorganisms (Jones & Hodges, 1974).

4. ENVIRONMENTAL LEVELS AND EXPOSURES

4.1 Environmental Levels

(a) Air

Atmospheric exposure to mirex could result from the air-borne dust from the production and processing of mirex or Dechlorane, combustion of either Dechlorane plastics or Dechlorane smoke compounds or volatilization of mirex used in bait formulations. The only information regarding any of the above occurrences is an estimate of potential volatilization of mirex based on the method of Gueckel et al. (1973).

(b) Water

Mirex has been found in one sample of ground water in the USA (Shackelford & Keith, 1976) and in water (0.0001 µg/litre) (Alley et al., 1973) shortly after bait application. Pond water in drainage areas is also known to contain high levels of mirex after treatment (0.2 and 0.53 µg/litre) (Spence & Markin, 1974). It has also been determined in rural drinking-water at levels of 0 - 437 ng/litre (Sandhu et al., 1978). However, mirex has not been found in tap water in studies with detection sensitivities as low as 5 ng/litre (Smillie et al., 1977).

(c) Food

The US tolerances for residues of mirex in food products in 1969 were: 0.1 mg/kg in all fat or meat of cattle, goats, horses, poultry and sheep, in milk fat, eggs, and fish; 0.01 mg/kg in all other raw agricultural commodities.

Mirex residues have been observed in beef fat in the southeastern USA and found to range between 0.001 mg/kg to 0.125 mg/kg with a mean of 0.026 mg/kg. No mirex was found in areas where bait was not used (Ford et al., 1973).

Plants are also a potential source for mirex uptake. Mirex residues of 0.01 - 1.71 mg/kg were found in soya beans, garden beans, sorghum, and wheat seedlings when grown on substrates containing 0.3 - 3.5 mg/kg mirex (de la Cruz & Rajanna, 1975). Based on these uptake data and the known soil concentrations of mirex, it has been calculated that plant tissues grown on contaminated soil could contain between 0.2 ng/kg and 2 µg/kg mirex (US EPA, 1978).

(d) Wildlife

Residue levels in various non-target organisms were wide-ranging and obviously dependent on the level of exposure and feeding habits. Also, as expected in all organisms, the highest levels of mirex were found in adipose tissue. Bird residue levels typically ranged from less than 1 mg/kg to 10 mg/kg. Residue concentrations of 210 mg/kg have been reported by Hallett et al. (1976) in lipids extracted from homing gulls from Lake Ontario.

Vertebrates such as frogs, lizards, and shrews have been observed to contain mirex residue levels as high as 9 mg/kg (Wojcik et al., 1975), 5.46 mg/kg (Markin et al., 1974b), and 41.3 mg/kg (Mirex Advisory Committee, 1972),[a] respectively. Again, typical residue levels are somewhat lower but generally range from approximately 1 to 10 mg/kg for the frog and lizard and 20 to 40 mg/kg in the shrew. It should be noted that these residue levels are maxima that are reached shortly after bait application and decrease over time. But, small amounts of mirex have been observed in tissues up to 3 years after application (Madhukar & Matsumura, 1979).

In areas where mirex has been detected in sediments or in the water, residue levels in aquatic animals have ranged from non-detectable to 0.97 mg/kg, with the majority of samples showing levels below 0.1 mg/kg. Fish in Lake Ontario and the St. Lawrence river contained levels as high as 0.27 mg/kg (Suta, 1978). For a more complete listing of US and Canadian residue data, see Baetcke et al. (1972), Borthwick et al. (1973, 1974), and US EPA (1978). World-wide data on mirex are lacking in the literature; however, mirex has been reported in Netherlands seals (Ten Noever de Brauw et al., 1973).

4.2 General Population Exposure

It has been estimated by the US EPA that inhabitants in areas treated with mirex bait would inhale some 0.4 - 0.8 ng mirex per day (Suta, 1978).

Food probably represents the major source of mirex accumulation in the human body. Within the food groups the largest intake of mirex would result from fish consumption, followed by wild game and then the commercial meats.

The average consumption of mirex via finfish would be 0.39 μg/day if the fish were from the St. Lawrence (US NRC, 1978). Mirex intake from Lake Ontario fish would on average be less than 0.34 μg/day. In the southern states,

---

[a] Report to US Environmental Protection Agency.

based on a mean mirex level of 0.02 mg/kg in southern fish, the average person would consume 0.13 µg of mirex per day. No data were available for mirex consumption internationally.

Wild game represents the second most significant mirex source. It has been estimated that in the USA approximately 9 million people will consume between 0.1 and 12 µg of mirex per person per day from wild game (US NRC, 1978).

(a) Infants

Mirex may be excreted in milk. A survey of 1436 samples of human milk, collected in the USA, failed to show detectable levels of mirex (Suta, 1978). However, in a Canadian survey, 3 out of 14 human milk samples showed levels between 2 - 21.5 µg/kg, on a fat basis (Mes et al., 1978).

(b) Occupational exposure

No data are available on occupational exposure to mirex.

5. KINETICS AND METABOLISM

5.1 Absorption

(a) Inhalation

Atallah & Dorough (1975) examined the transfer of mirex from cigarettes, through cigarette smoke, using ^{14}C-mirex in rats. Of the total residue inhaled, 47% was exhaled, the lung retaining 35%, the blood 11%, and heart 1%, 2 - 4 min after inhalation.

(b) Gastrointestinal tract

Approximately 55% of a single oral dose of 6 mg ^{14}C-mirex/kg body weight, administered to rats, was excreted unchanged in the faeces within 48 h (Mehendale et al., 1972). When a lower concentration of mirex was administered (0.2 mg/kg body weight), only 15% of the administered mirex was excreted in 48 h (Gibson et al., 1972). Ivie et al. (1974b) dosed Japanese quail orally with 1.2 mg ^{14}C-mirex/kg body weight and found that only 12 - 25% of the dose was eliminated in the faeces after 1 week. In another study, a female rhesus monkey was given ^{14}C-mirex orally at 1 mg/kg body weight. ^{14}C-mirex appeared in the plasma after 2 h and reached a peak after 5 h (Wiener et al., 1976).

(c) Skin

No studies on dermal uptake were found.

5.2 Distribution and Storage

(a) Human studies

The first discovery of mirex residues in human adipose tissue was reported by Kutz et al. (1974). The levels found in 6 post-mortem samples, all from patients who had resided in the southeastern states of the USA, ranged from 0.16 to 5.94 mg/kg. More recently, the US EPA reported that 18% of 284 samples obtained from the southeast area general population contained mirex and that values ranged from trace amounts to 1.32 mg/kg (Suta, 1978). Lloyd et al. (1974) analysed the blood of pregnant women in the Jackson and Mississippi delta areas for chlorinated pesticides including mirex. Mirex was found in 106 of the 142 samples of this survey at a mean blood concentration of 0.5 µg/litre.

(b) Animal studies

Mirex is a lipophilic compound and as such is stored in the adipose tissue to a much greater extent than in any other tissue. Mehendale et al. (1972) showed that when rats were dosed with a single oral dose of mirex at 6 mg/kg body weight, the tissues and organs retained about 34% of the total dose, of which 28% was found in fat, 3.2% in muscle, 0.09% in the kidneys, and 1.8% in the liver.

Ivie et al. (1974b) reported on the accumulation, distribution, and excretion of ^{14}C-mirex fed to rats and quail for 16 months at levels of 0.3, 3, or 30 mg/kg diet. The levels of mirex in the fat of rats and quail were about 120 to 185-fold greater than the dietary intake values, and no plateau was observed in the accumulation pattern. As part of this study, rats and quail were given mirex-treated food for 6 months and then placed on a control diet for an additional 10 months. Analyses of tissues indicated that the half-life of mirex in the quail was 20 - 30 days, whereas, in the rat, the residues had declined by only 40% after 10 months.

The distribution of mirex in female rhesus monkeys, dosed orally and intravenously at approximately 1 mg/kg body weight, was studied by Wiener et al. (1976). Peak concentrations in the plasma were of the order of 1 mg/litre in the iv-treated animals and approximately 0.01 mg/litre, 400 days later. At autopsy, all tissues examined contained mirex. In a reproduction study on rats, mirex was transferred to the fetus across the placenta and was also excreted in the milk (Gaines & Kimbrough, 1970). Rats fed mirex at 25 mg/kg diet for 78 days excreted 11.3 mg/litre in milk, whereas fetuses removed by Caesarian section on the 19th day of gestation contained 0.23 mg/kg body weight.

Distribution studies have also been reported on the cow (Bond et al., 1975), goat (Smrek et al., 1977, 1978), mosquito fish (Ivie et al., 1974b), wild birds (Stickel et al., 1973), blue crab (Schoor, 1974), and winter flounder (Pritchard et al., 1973).

5.3 Metabolism

Mirex does not appear to be metabolized to any significant extent in any animal species so far investigated (mice, rats, rabbits, monkeys) (Waters, 1976; Canada, Department of National Health and Welfare, 1977; IARC, 1979).

5.4 Excretion

(a) Animal studies

Data from studies on rat (Gibson et al., 1972; Mehendale et al., 1972; Ivie et al., 1974b), monkey (Wiener et al., 1976), quail (Kendall et al., 1978), and goat (Smrek et al., 1977), exposed to mirex, showed fast tissue uptake and slow elimination. Mehendale et al. (1972) estimated the half-life of mirex following oral administration to rats to be more than 100 days. Pittman et al. (1976) used a mathematical model to predict an extremely long half-life for mirex in rhesus monkeys with only a 2% decline in adipose tissue levels over a 10-year period. However, after a 52-week recovery period, the mirex level in the adipose tissue of goats was one-third to one-quarter of the original value (Smrek et al., 1978). In a feeding study on rats, quail, and mosquito fish (Ivie et al., 1974b), a 40% decline in mirex levels in adipose tissue was found over a 10-month period, while the half-life of mirex was 20 - 30 days in the adipose tissue of quails and 4 months in fish. In rats, 12 - 25% of the dose was eliminated in the faeces after 1 week (Ivie et al., 1974b).

6. EFFECTS ON EXPERIMENTAL ANIMALS

6.1 Single-Dose Studies

Data indicating the acute oral, intraperitoneal, and dermal toxicity for mirex in various animals are shown in Table 2. The acute toxic effects of mirex were characterized by muscle tremors, diarrhoea, and depression followed by death (Gaines & Kimbrough, 1970).

Table 2. Oral, intraperitoneal, and dermal LD_{50} values for mirex

Species	Sex	Exposure route	LD_{50} (mg/kg body weight)	Reference
Rat	M	oral (corn oil)	740	Gaines (1969)
Rat	F	oral (corn oil)	600	Gaines (1969)
Rat	M & F	oral (peanut oil)	3000	Gaines (1969)
Rat	F	oral (corn oil)	365	Gaines & Kimbrough (1970)
Hamster	F	oral	125	Cabral et al. (1979)
Hamster	M	oral	250	Cabral et al. (1979)
Dog	M	oral (corn oil)	1000	Larson et al. (1979)
Rat	F	ip	365	Kendall (1974)
Rabbit	-	dermal	800	Waters (1976)
Rat	M & F	dermal	2000	Gaines (1969)

Several hepatic variables were studied, 2 days following a single oral dose of 100 mg mirex/kg body weight, in female rats. Microsomal cytochrome P-450 content, NADPH-cytochrome _c_ reductase (EC 1.6.2.4) activity, and hepatic ascorbic acid concentration were found to be increased, and so was the microsomal protein concentration. Relative liver weight was increased, as well as the activities of aminopyrine _N_-demethylase and 4-nitroanisole-_O_-demethylase (Chambers & Trevathan, 1983).

6.2 Short-Term Studies

6.2.1 Oral exposure

The toxic effects of mirex in short-term studies are generally characterized by a decrease in body weight, hepatomegaly, induction of mixed-function oxidases, morphological changes in liver cells, and sometimes death.

Decreased body weight gain was observed in female rats fed a total of 365 mg/kg body weight over a 12-day period (Kendall, 1974), and in male rats dosed orally for 14 days at 10 mg/kg (Villeneuve et al., 1977). In a 13-week feeding study, decreased body weight gain was observed in female rats at a dietary level of 1280 mg/kg and in male rats at 320 and 1280 mg/kg (Larson et al., 1979). Reduced body weight gain was also observed when beagle dogs were fed mirex at 100 mg/kg diet for 13 weeks (Larson et al., 1979).

Liver hypertrophy was observed in: male rats dosed by gavage with 1.0 and 10 mg mirex/kg body weight, in corn oil, for 14 days (Villeneuve et al., 1977); male and female rats dosed from 5 - 50 mg/kg body weight for 5 days (Mehendale et al., 1973); male and female rats dosed orally once, with 50 mg mirex/kg body weight and then observed for 28 days (Robinson & Yarbrough, 1968); and in male rats dosed ip with 50 mg/kg body weight for 5 days (Kaminsky et al., 1978). Female rats fed 20, 30, or 40 mg mirex/kg and male rats fed 40 or 50 mg mirex/kg for 28 days exhibited liver enlargement (Abston & Yarbrough, 1976).

In another study, rats (sex not specified) fed 100 mg mirex/kg diet for 4 weeks, exhibited liver hypertrophy (Davison et al., 1976), whereas in a study carried out over 166 days, liver hypertrophy was observed at 25 mg/kg diet for both sexes (Gaines & Kimbrough, 1970). When mirex was fed for 13 weeks to male and female rats, liver hypertrophy was observed at levels of 80 mg/kg and higher in males and at 320 mg/kg in females (Larson et al., 1979). Liver enlargement was also observed in female rabbits fed 20 mg mirex/kg diet for 8 weeks (Warren et al., 1978), in dogs fed 100 mg/kg for 13 weeks (Larson et al., 1979), and in male mice fed 30 mg/kg for 12 weeks (Pitz et al., 1979).

Induction of mixed-function oxidase (EC 1.14.14.1) enzymes was shown for the male rat, when mirex was administered: by gavage at levels as low as 1.0 mg/kg body weight per day, for 14 days (Villeneuve et al., 1977); at 5 mg/kg per day ip for 5 days (Kaminsky et al., 1978); at 5 mg/kg diet (0.5 mg/kg per day) for 13 weeks (Villeneuve et al., 1979b); and at 1 mg/kg diet for 14 days (Iverson, 1976). Mirex has also been shown to induce microsomal enzyme activity in rabbits when administered at 20 mg/kg diet for 8 weeks (Warren et al., 1978) in neonatal mice, suckled on mothers exposed to 10 mg mirex/kg diet (Fabacher & Hodgson, 1976), but not in chickens or quail exposed to 160 or 80 mg mirex/kg diet for 16 and 12 weeks, respectively (Davison et al., 1976). In a study designed to investigate the type of enzyme induction, mirex was found to induce a pattern similar to phenobarbital, DDT, chlordane, and chlordecone (Madhukar & Matsumura, 1979).

Morphological changes observed in the liver of mirex-treated rats consisted of hepatocyte enlargement, depletion of glycogen and lipid accumulation (Kendall, 1974, 1979), and some cell necrosis (Kendall, 1974; Davison et al., 1976). Ultrastructural changes included altered architecture of the rough endoplasmic reticulum (RER), dilated RER cisternae, an increase in the number of free ribosomes, and proliferation of the smooth endoplasmic reticulum (Gaines & Kimbrough, 1970; Kendall, 1979). The lowest level reported to cause histological changes was 1.0 mg/kg diet, and was observed in male rats during a 166-day study (Gaines & Kimbrough, 1970) and a 90-day study (Villeneuve et al., 1979c).

Other important effects observed in several studies were the mirex-induced impairment of hepatobiliary function (Mehendale, 1976, 1977a, 1979; Mehendale et al., 1979), and bile stasis (Gaines & Kimbrough, 1970).

6.2.2 Dermal exposure

In a short-term dermal study (Larson et al., 1979), rabbits were exposed to 3.33 or 6.7 g of mirex bait/kg body weight for 6 - 7 h each day, 5 days a week, for 9 weeks. There were no gross or histopathological changes resulting from treatment in any of the animals.

6.3 Long-Term and Carcinogenicity Studies

The long-term and carcinogenic effects of mirex are summarized in Table 3. Some of these studies have been discussed extensively by IARC (1979). The data indicate that mirex is carcinogenic for rats and mice (IARC, 1979).

6.4 Reproduction and Teratogenicity Studies

Ware & Good (1967) carried out a study on mice and found that administration of 5 mg mirex/kg diet for 30 days, prior to mating, resulted in a reduced litter size. In a more recent study, Wolfe et al. (1979) found a cessation in reproduction in mice fed 17.8 mg/kg diet for 3 months and decreased reproduction in the group fed 1.8 mg/kg. Gaines & Kimbrough (1970) fed diets containing 25 mg mirex/kg to rats and found reduced litter size, reduced viability of the neonates, and cataract formation in surviving neonates. In addition, the results of a cross-fostering study indicated that the formation of cataracts was due to exposure through the milk. Females fed mirex at 5 mg/kg produced normal litters.

Pregnant rats were given 6 mg mirex/kg body weight per day in an oily solution, by gavage, on days 8 1/2 - 15 1/2 of

Table 3. Summary of long-term and carcinogenicity studies with mirex

Species	Duration	Doses used	Effects	Reference
Mouse	up to 70 weeks	1 - 90 mg/kg diet	increased liver weights at 5 mg/kg & higher, mixed function oxidase activity increased at 1 mg/kg after 70 weeks; total liver DNA & total liver protein & mitochondrial respiration increased at 1 mg/kg, after 70 weeks	Byard et al. (1975)
Rat	up to 36 months	5 and 30 mg/kg diet	no effects on liver weight; proliferation of SER observed after 12 months at both dose levels	Fulfs et al. (1977)
Mouse (2 strains)	70 weeks	dosed orally with 10 mg mirex/kg from day 7-28 after birth, then placed on a diet containing 26 mg mirex/kg until 70 weeks of age	increased incidence of hepatomas in both strains	Innes et al. (1969)
Mouse	78 weeks	mice received 1 single subcutaneous injection of 1000 mg mirex/kg body weight in gelatine on their 28th day of life	increased incidence of reticulum-cell sarcomas	US NTIS (1968)

Table 3 (contd).

Species	Duration	Doses used	Effects	Reference
Mouse	up to 18 months	1, 5, 15, and 30 mg/kg diet	increased liver weights at 1 mg/kg in female mice, 5 mg/kg and higher in male mice; histological changes at 5 mg/kg and higher; proliferation of SER observed ultrastructurally at 1 mg/kg and above	Fulfs et al. (1977)
Monkey	up to 26 months	0.25 and 1.0 mg/kg body weight orally 6 days per week (equivalent to 5 and 20 mg/kg in diet)	no effect on liver weights, liver histology, or liver ultrastructure	Fulfs et al. (1977)
Rat	18 months	50 or 100 mg/kg diet exposure, + 6 months on control diet	dose-related effect on survival noted; increased incidence of neoplastic nodules observed in high-dose male rats; of 17 rats from all groups, 6 animals including 4 high-dose males had liver-cell carcinomas; no metastases were observed	Ulland et al. (1977)

pregnancy. The majority of the moderate to severely oedematous fetuses had abnormal ECGs and were either dead or dying on the morning before parturition was expected (Grabowski & Payne, 1983).

In reproduction studies on birds, dietary administration of mirex did not reduce egg production or embryo survival in chickens (Davison & Cox, 1974), mallards, or bobwhite quail (Heath & Spann, 1973).

The teratogenic potential of mirex was studied in rats given daily oral doses of 0, 1.5, 3.0, 6.0, or 12.5 mg/kg body weight on days 6 - 15 of gestation (Khera et al., 1976). The 12.5 mg/kg dosage caused maternal toxic effects, decreased fetal survival, reduced fetal weight, and an increased incidence of visceral anomalies in the fetus. Maternal effects and increased incidence of visceral anomalies in the fetus were observed at 6.0 mg/kg body weight. The lower doses did not induce any adverse effects.

Mirex administered to pregnant rats at 7 mg/kg body weight per day during days 7 - 16 of gestation, and also post-partum, induced oedema, undescended testes, and reduced weight in offspring (Chernoff et al., 1976). Mirex-induced cataract formation was observed in mice in the same study.

6.5 Mutagenicity

Mirex was negative in a dominant lethal test on rats, in which doses of 1.5 - 6.0 mg/kg body weight per day were used (Khera et al., 1976). Mirex was found to be negative when tested by the standard Ames bacterial assay including a liver microsomal activation mixture (Hallett et al., 1978).

6.6 Other Studies

Adult male rats were fed diets containing mirex at 1.78 and 17.8 mg/kg for several weeks and were tested on a variety of behavioural tasks. No differences in behaviour were seen between control and treated animals (Thorne et al., 1978). In a study in which male rats were fed diets containing mirex at levels up to 80 mg/kg diet for 8 weeks, mirex was found to cause hyporeactivity with attenuated startle response, increased emergence time, and decreased ambulation (Reiter et al., 1977). The results of other studies also indicated that mirex might influence behaviour (Peeler, 1976; Reiter et al., 1977; Dietz & McMillan, 1978).

Studies have been conducted to evaluate the influence of mirex on antibody-mediated immunity in the chicken (Glick, 1974). A level of 500 mg mirex/kg diet for up to 5 weeks of age depressed the levels of immunoglobulin M and G but did not affect antibody production. The same observations were made in a subsequent study (Rao & Glick, 1977) where chickens were fed diets containing mirex at 100 mg/kg for 40 days from hatching.

Pregnant mare serum (PMS)-induced ovulation was significantly inhibited in immature rats by a single administration of 0.4 - 50 mg of mirex per animal (Fuller & Draper, 1975). This suppression of PMS-induced ovulation was thought to be due to an action on the central nervous system, inhibiting the release of luteinizing hormone rather than to a direct effect on the ovary.

The effects of mirex on Ehrlich ascites tumour cells have been assessed using certain *in vivo* and *in vitro* measurements (Walker et al., 1977). Mirex retarded the development of this tumour *in vivo* and inhibited the synthesis of RNA purines.

Food deprivation has been shown to enhance the inducing properties of mirex on the mixed function oxidases (Villeneuve et al., 1977).

Both the acute toxicity data (Table 2) and some short-term exposure effects (Larson et al., 1979) seem to suggest that the female is more sensitive to mirex than the male, but no study has addressed this question specifically.

Several publications include reports on the toxicological properties of the mirex breakdown products *photomirex* (8-monohydromirex) and 2,8-dihydromirex. The results of short-term studies indicate that photomirex can induce: liver enlargement, mixed-function oxidases, histological changes in the liver, thyroid, and testes, and even death (Villeneuve et al., 1979b,c; Sundaram et al., 1980). The histological changes induced by photomirex (the most sensitive of the variables mentioned) generally occurred at levels approximately one order of magnitude lower than those observed with mirex. Dihydromirex also causes morphological changes in the liver and thyroid but generally at the same dosage levels as mirex (Chu et al., 1980a). Photomirex is not teratogenic in the rabbit, but does cause reproductive impairment in the rat including cataract formation in the pups (Villeneuve et al., 1979a; Chu et al., 1981). Photomirex has a very long half-life in primates and only 10% of the administered dose was eliminated over a one-year period (Chu et al., 1982).

Ultrastructural changes in the thyroid follicular cells of male rats persisted for at least 18 months following cessation of a 28-day exposure to 0.05 - 50 mg photomirex/kg diet or 50 mg mirex/kg diet. The morphological changes consisted of increased follicular cell heights and a numerical increase in secondary lysosomes. In the 50 mg mirex/kg group, columnar thyroid follicular cells were engorged with deformed lysosomal bodies (Singh et al., 1982).

7. EFFECTS ON MAN

No reports of poisoning incidents or levels of occupational exposure are available.

8. EFFECTS ON ORGANISMS IN THE ENVIRONMENT

8.1 Aquatic Organisms

Information on the toxicity of mirex is available for a wide range of aquatic organisms.

Data on mirex toxicity for a variety of algae are given in Table 4. A more comprehensive table, listing different conditions and exposure times is available on request from the IRPTC, Geneva. Results of disc assay tests of estuarine bacterial growth inhibition were inconsistent from batch to batch of technical grade mirex, with some batches producing little or no growth inhibition in the bacterial isolates while others showed marked inhibition (Brown et al., 1975). Although purified mirex was not toxic, UV-irradiated mirex was bacteriologically toxic. The only appreciable microbial activity affected by mirex at concentrations below 100 mg/litre was the inhibition of primary production. This is unlikely to be a significant effect in the field, since most phytoplankton are in the aqueous phase, whereas mirex tends to become associated with the sediments. Mirex degradation products with substitution at the 5 and/or 10 positions were highly toxic for bacterial cultures and, as these compounds are more polar than mirex, they may be more soluble in water and therefore pose a greater environmental threat for aquatic bacteria.

Exposure of phytoplankton to mirex at 1 mg/litre for 4 h reduced productivity by 28 - 46% (Butler, 1963). The ciliate protozoan *Tetrahymena pyriformis* exhibited reduced growth rate when exposed to 0.9 µg mirex/litre during the exponential growth phase (Cooley et al., 1972). Exposure of pure cultures of the green marine algae *Chlamydomonas* sp. to 1 mg mirex/litre for 168 h reduced net photosynthesis by 55% and respiration by 28.4% (de la Cruz & Naqvi, 1973). Population growth and oxygen evolution in marine unicellular algae were not affected by exposure to 0.2 µg mirex/litre (highest concentration of mirex obtainable in seawater) when tested under various conditions of salinity and nutrient concentration (Hollister et al., 1975). Exposure to 10.2 µg/litre (maximum concentration of mirex obtainable in synthetic seawater) did not adversely affect photosynthesis and the chemical composition of green and red marine algae (Sikka et al., 1976). Mirex at a concentration of 100 µg/litre in a culture medium of a freshwater algae, *Chlorella pyrenoidosa*, depressed population growth by 8% in 92 h and 19% in 164 h (Kritcher et al., 1975).

Mirex is highly toxic for crustacea; data are summarized in Table 5. Delayed mortality appears to be characteristic of

Table 4. Toxicity of mirex for algae

Organism	End point	Parameter	Concentration (µg/litre)	Reference
Alga	decrease in productivity	4-h EC_{28-46}	1000	Butler (1963)
Algae:				
Uva lactuca *Enteromorpha linza* *Rhodymenia pseudopalmata*	no effect at maximum solubility in water	EC_0	10.2	Sikka et al. (1976)
Marine algae:				
Chlorococcum sp. *Dunaliella tertiolecta* *Chlamydomonas* sp. *Porphiridium cruentum* *Thallasiosira pseudonana* *Nitzschia* sp.	no effect on population growth or oxygen evolution	168-h EC_0	0.2	Hollister et al. (1975)

Table 5. Toxicity of mirex for crustacea

Organism	Size	Flow/stat	pH	Temp (°C)	Hardness (mg/litre)	Salinity (°/oo)	Parameter	Concentration (µg/litre)	Reference
Crayfish, juvenile (*Procambarus hayi*)	0.6 cm	aerated stat	7.8		28		48-h LC_{65}	0.1	Ludke et al. (1971)
Blue crab, larva (*Callinectes sapidus*)		stat		25		30	20-day 98% survival to megalopa	1	Bookhout & Costlow (1976)
		stat		25		30	20-day 0.5% survival to 1st crab	1	Bookhout & Costlow (1976)
		stat		25		30	20-day 56% survival to megalopa	0.01	Bookhout & Costlow (1976)
		stat		25		30	20-day 41.5% survival to 1st crab	0.01	Bookhout & Costlow (1976)

Table 5 (contd).

Organism	Size	Flow/stat	pH	Temp (°C)	Hardness (mg/litre)	Salinity (°/oo)	Parameter	Concentration (µg/litre)	Reference
Blue crab, juvenile (*Callinectes sapidus*)		flow 400 l/h					96-h exposure 100% died within 18 days	100	Lowe et al. (1970)
Pink shrimp, juvenile (*Penaeus duorarum*)	51-76 mm	flow 400 l/h		14		29	3-week LC_{11}	0.1	Lowe et al. (1971)
Glass shrimp (*Palaemonetes kadiakensis*)		stat		23-25			120-h LC_{50}	190	Naqvi & de la Cruz (1973b)
Amphipod (*Hyallela azteca*)		stat		23-25			600-h LC_{54}	1	Naqvi & de la Cruz (1973b)

mirex poisoning in crustacea. Freshwater crayfish, particularly third instars, were extremely sensitive to mirex, through direct and indirect exposure under laboratory conditions (Ludke et al., 1971). Although authors have suggested that crayfish would not be exposed to sufficient mirex under field conditions to cause population decline (Muncy & Oliver, 1963; Markin et al., 1972), three applications of mirex bait at 1.4 kg/ha, about 90 days apart, reduced the number of red crayfish eventually harvested (Hyde, 1973). Estuarine crustaceans exposed to mirex under laboratory conditions become irritated, lose equilibrium, move randomly, become paralysed and may die (Bell et al., 1978). The onset and severity of such symptoms depend on the level of exposure, water temperature (Tagatz et al., 1975) and salinity (Leffler, 1975), and the age and size of animal under test (Lowe et al., 1971). Juvenile and larval stages are most sensitive. Exposure to 0.01 - 10 µg mirex/litre medium did not have any appreciable effect on day-to-day survival in 2 replicate series of larval blue crabs, for 5 days after hatching. Delayed mortality then occurred with 1 and 10 µg being acutely toxic and 0.01 and 0.1 being sublethal (Bookhout & Costlow, 1976). McKenzie (1970) found that the toxicity of mirex bait for crabs was temperature dependent; no mortality occurred in treated crabs held at 10 °C but survival time decreased as the water temperature increased from 20 to 27 °C.

At subacute internal levels of mirex (0.19 - 0.03 mg/kg body weight), caused by the ingestion of 0.14 µg mirex bait, crabs held in water of intermediate salinity (6.8 - 20.4°/oo) showed an elevated metabolic rate, inhibition of limb autotomization, thin carapaces, and abnormal behaviour. A sub-acute level of 0.01 µg/litre medium lengthened the duration of the developmental stages in mud crabs but had no effect on stone crab development (Bookhout et al., 1972). In a simulated field application of mirex fire ant bait, the bait was applied at 1.4 kg/ha on a sandy slope with a pool of flowing seawater (29 °C; salinity 27°/oo) at the other end of a tank. After 2 treatments, one week apart, 73% of fiddler crabs became paralysed or died within 2 weeks of the applications (Lowe et al., 1971).

Oxygen consumption of the pond snail, Physa gyrina was increased by exposure to low concentrations of mirex (0.008 - 0.07 mg/litre) for 3 days but decreased by 44% by exposure to 1 mg/litre mirex (de la Cruz & Naqvi, 1973). Exposure to very low concentrations of mirex (initial concentration 0.062, final concentration of 0.016 µg/litre) for 30 days was sufficient to decrease the feeding and burrowing activities of adult lugworms, Arenicola cristata, even 45 days after termination of the exposure (Schoor & Newman, 1976).

Some toxicity data for fish are given in Table 6. Young juveniles are more sensitive to mirex than adults (Lee et al., 1975).

Bluegills and goldfish were exposed to mirex either through a single application of a formulation to holding ponds or by mirex-treated diet (Van Valin et al., 1968). No mortality occurred in bluegills, but growth was slow in fish fed 5 mg/kg diet. A dose-related mortality rate and pathological changes were observed in goldfish exposed to 0.1 and 1 mg/litre. Goldfish that died at these concentrations were emaciated, lacked slime layers, had roughened skin with many protruding scales, and exhibited oedematous gill changes. Survivors suffered microbial infection (the severity of which was related to treatment level), distended gall bladders, and granulomatous kidney lesions by day 224. In another study, treatment of ponds with mirex reduced the survival of 4 species of fish to 43.3%, compared with 71.6% survival in controls, 10 months after the application, but spawning was not affected (Bookhout & Costlow, 1975). Mosquito fish and bluegills exposed to 1 mg mirex/litre leached from fire ant bait exhibited differences in oxygen consumption compared with controls during the 7-week exposure, but whether the differences were significant was not stated (de la Cruz & Naqvi, 1973). Mirex affects fish behaviour. Temperature selection was altered in sailfin molly exposed to mirex at 1 mg/litre (Degrove cited in Task Force Report on Mirex, 1977). The activity rhythm of diamond killifish was affected by exposure to mirex (Tolman & Livingston cited in Canada, Department of National Health and Welfare, 1977).

8.2 Terrestrial Organisms

8.2.1 Plants

Little work has been done on the effects of mirex on terrestrial plants. In one study (Rajanna & de la Cruz, 1975), the phytotoxic effects of recrystallized technical mirex on 6 crops were investigated. Reduction in germination and emergence occurred as the concentration of mirex increased. In germination studies, where germination blotters were soaked in solutions of mirex, the percentage of germination occurring over 21 days was significantly reduced by 0.15 mg/litre in tall fescue, alisike clover, and alfalfa; by 0.3 mg/litre in crimson clover and johnson grass; and by 0.7 mg/litre in annual rye grass. Similar doses caused a reduction in percent emergence, when mirex was applied to the sandy substrate in which the seeds were grown. In a duplicate study, emerged seedlings were harvested 2 weeks after planting, for dry weight determinations. Significant

Table 6. Toxicity of mirex for fish

Species	Flow/ stat	Temp (°C)	Parameter	Concentration (μg/litre)	Effect	Reference
Goldfish (*Carassius auratus*)	stat	‘2-28		1000	death 24 days after exposure 75-100%, granulomatous lesions of kidney 224 days after exposure	Van Valin et al. (1968)
Mullet (*Mugil cephalus*)	flow 15.61 1/h		96-h	10-10 000	adult (260-380 mm), old juv. (70-150 mm); no deaths; mortality rate in young juv. (20-43 mm); 6.4% (0.01 mg/litre), 26.9% (0.1 mg/litre), 32.1% (1 mg/litre), 90% (10 mg/litre)	Lee et al. (1975)
Bluegill (*Lepomis macrochirus*)			LC_0	1.3-1000	no effect on reproduction, body weight, size; population decreased	Van Valin et al. (1968)
Sheepshead minnow (*Cyprinodon variegatus*)[a,d]	flow	19.1 29.8		0.53	some gill changes	Tagatz et al. (1975)
Pinfish, juvenile (*Lagodon rhomboides*)[b]	flow		5-month	20 mg/kg diet	no effect	Lowe et al. (1971)
Channel catfish (*Ictalurus punctatus*)[c]	stat			1.4 kg/ha 3 applications	survival decreased by 39.5% compared with control	Hyde et al. (1974)

[a] Exposed to leachate from fire ant bait.
[b] Technical grade, 98% mirex.
[c] Three applications of mirex bait 0.3% technical.
[d] Four 28-day seasonal exposures.

reductions in growth rate occurred at 0.15 mg/litre in crimson clover, johnson grass and annual rye grass; at 0.3 mg/litre in fescue and alfalfa; and at 0.7 mg/litre in alisike clover. Visual examination of seedlings revealed poor development. Other studies have demonstrated uptake, accumulation (de la Cruz & Rajanna, 1975), and translocation (Mehendale et al., 1972) of mirex in plants, but there was no evidence of metabolic transformation.

8.2.2 Insects

Mirex is moderately toxic for bees and should not be applied directly to bees in the field or in the colonies; the authors quote an LD_{50} of 7.15 µg/bee (Atkins et al., 1975). In adult field crickets, a lethal dose of 25 µg per animal produced characteristic symptoms of a latent period of at least 72 h followed by hyperactivity, ataxia, convulsions, and paralysis (MacFarlane et al., 1975). The primary action of mirex was suggested to be on synaptic transmission; mirex exposure results in prolonged synaptic after discharge and enhanced spontaneous transmission. Houseflies respond slowly to mirex (Plapp, 1973). Exposure of 11 strains to 1 mg mirex residues per jar of flies resulted in a 50% knock-down in 2 - 4 days exposure and 90% knock-down in 3 - 5 days. Exposure to 100 mg/kg diet produced 50% knock-down in 4.5 - 7 days and 90% in 5 - 10 days. With lower concentrations, knock-down time increased and differences in response between strains became greater.

8.2.3 Birds

Mirex is not very toxic for birds (Table 7). It is of low short-term toxicity for wild birds; dietary doses of 2250, 750, and 250 mg/kg diet killed 50% of juvenile male grackles in 5, 14, and 38 days, respectively. Death occurred sooner in colder weather, presumably because food consumption increased (Stickel et al., 1973).

Most toxicological studies on birds have monitored the effects of mirex on reproductive variables. Under field conditions, no significantly adverse effects on reproduction were observed when bobwhite quail were kept on plots treated with 11.2 (the regular field-rate use), 112, or 1120 kg/ha (Baker, 1963). In laboratory studies, feeding bobwhite quail with 40 mg mirex/kg, and mallard with 1 or 10 mg/kg in their diet, did not have any effects on egg production, shell strength or thickness, embryonation and embryo survival, or hatching and survival of chicks up to 14 days of age (Heath & Spann, 1973). Exposure of white leghorn hens to 5, 10, 20, 80, and 160 mg mirex/kg diet, and of Japanese quail to 5, 40,

Table 7. Toxicity of mirex for birds

Species	Route	Age	Sex	Parameter	Concentration (mg/kg)	Reference
Mallard	oral			acute LD_{50}	2400[a]	Waters (1976)
Mallard	oral	10 day		8-day LD_{50} [b]	> 5000	Hill et al. (1975)
Mallard	oral	3-4 month	M	acute LD_{50} [c]	2400[a]	Tucker & Crabtree (1970)
Japanese quail	oral			acute LD_{50}	10 000	Waters (1976)
Japanese quail	oral	14 day		8-day LD_{50} [b]	> 5000	Heath et al. (1972)
Bobwhite quail	oral	14 day		8-day LD_{50} [b]	2511	Heath et al. (1972)
Pheasant	oral			acute LD_{50}	1400-1600	Waters (1976)
Ring-necked pheasant	oral	14 day		8-day LD_{50} [b]	1540	Heath et al. (1972)
Grackle	oral	juv.	M	12-day LD_{50}	750	Stickel et al. (1973)
Cowbird	oral	adult	M	12-day LD_{50}	750	Stickel et al. (1973)
Redwinged blackbird	oral	adult	F	11-day LD_{50}	750	Stickel et al. (1973)
Starling	oral	juv.	F	9-day LD_{50}	750	Stickel et al. (1973)

[a] mg/kg body weight, otherwise mg/kg diet.
[b] Fed mirex for 5 days, untreated diet for 3 days, mortality estimated on day 8.
[c] Single dose, mortality estimated 14 day post treatment.

and 80 mg/kg, for 12 weeks did not affect egg production, egg weight, shell calcium content, shell thickness, shell weight, or the proportion of broken or soft-shelled eggs (Davison & Cox, 1974; Davison et al., 1975). Laying hens tolerated up to 200 mg mirex/kg diet without adverse effects on hatchability or chick growth and survival, but there were some eggshell abnormalities (Waters, 1976). Twelve weeks of daily exposure to mirex at 300 or 600 mg/kg in the feed produced weight loss in hens. Exposure to 600 mg/kg also caused a significant decrease in egg hatchability and chick survival (Naber & Ware, 1965). Exposure of third-generation progeny of wild mallards to a diet treated with mirex at 100 mg/kg for 25 weeks caused a significant reduction in duckling survival (Hyde et al., 1973b). The percentage of ducklings surviving up to 2 weeks after hatching was 72.6 in the 100 mg/kg group compared with 93.8 and 95.7 in the 1 mg/kg and control groups, respectively. There appeared to be a deleterious association between residue concentration in the egg and subsequent duckling survival. When 0.1 mg mirex/kg diet was fed to laying white leghorn hens in combination with similar low levels of dieldrin, DDT, and heptachlor, there were no synergistic effects on reproductive variables (Driver et al., 1976).

In studies on the biochemical effects of mirex on birds, dietary levels of 5 - 80 mg/kg, fed to quail, and 5 - 160 mg/kg fed to chickens, for 12 weeks, did not affect liver weight, aniline hydroxylase and aminopyrine-N-demethylase activities of hepatic microsomes, or cytochrome P-450 concentrations in hepatic microsomes (Davison & Cox, 1974). Chickens fed 10 or more mg mirex/kg diet showed structural changes in their livers; 500 mg/kg fed to newly hatched chickens up to 5 weeks of age significantly depressed levels of IgG and IgM but did not influence antibody production (Glick, 1974).

8.3 Microorganisms

Estuarine microorganisms are not affected by concentrations of mirex that are likely to be found in the estuarine environment. The only variable affected by mirex at concentrations below 100 mg/litre is primary productivity (Brown et al., 1975). However, mirex is rapidly associated with sediments and the highest concentrations recorded in Lake Ontario were around 40 μg/litre (Canada, Department of National Health and Welfare, 1977). The degradation products of mirex, e.g., kepone and photomirex, are more toxic than the parent compound (Brown et al., 1975).

Total populations of soil fungi and bacteria were not affected by exposure to 20 g technical mirex/kg soil for 7 days, but concentrations of 10 and 20 g/kg did reduce the actinomycete population in 1 out of 3 soils treated (Jones & Hodges, 1974).

8.4 Bioaccumulation and Biomagnification

Mirex is highly cumulative; bioaccumulation data are summarized in Table 8. The amount taken up varies with species, and is also related to the concentration and duration of exposure (de la Cruz & Naqvi, 1973). Kobylinski & Livingston (1975) studied the uptake of mirex from contaminated sediment by the Hogchoker (a freshwater flatfish) under both static and constant flow conditions. Mirex was added to sand at concentrations of 1430, 470, and 140 μg/kg, and this was covered by 14 litres of water. Uptake by Hogchoker tissues was dose-dependent, with accumulation increasing over time without reaching an equilibrium. Fish absorbed mirex from both the water and sediments. In the flowing system, appreciable amounts of mirex were lost from the environment.

Wojcik et al. (1975) measured mirex residues in a variety of non-target organisms and suggested that biomagnification had occurred. This had presumably followed ingestion of animals containing lower residues. Residues tended to be higher in insectivorous species that ate targeted insects. In fish, residues tended to be higher in predators than in omnivores (Collins et al., 1974). It has been shown that mirex can be moved through a simple 2-level food chain by feeding crab on shrimp that had been poisoned by mirex (Lowe et al., 1971). During mirex treatment of coastal areas for fire ant control and for a year afterwards, residues were less than 10 μg/litre in water and were 0 - 0.07 mg/kg in sediment (Borthwick et al., 1973). However, in organisms up the food chain, concentrations increased significantly; birds contained 0 - 0.17 mg/kg and mammals 0 - 4.4 mg/kg.

In birds, mirex may accumulate to high levels. This is particularly so in insectivorous birds where mirex levels of 1 - 10 mg/kg tissue have been reported (Mirex Advisory Committee, 1972).[a] Mirex residues transferred to eggs persist in juveniles. For example, in snowy egrets, eggs contained 13 mg/kg, nestlings 3 - 5 mg/kg, and parents, 0.64 mg/kg, after an application of mirex baits in the area (Mirex

---

[a] Report to US Environmental Protection Agency

Table 8. Bioaccumulation of mirex

Organism	Organ	BCF	Concentration (µg/litre)	Exposure period	Conditions	Reference
Turtle grass (*Thalassia testudinum*)	leaf rhizome	0 0.36	0.1	10 days	exposed through rhizomes	US EPA (1972)
Ulva lactuca *Enteromorpha linza* *Rhodymenia pseudopalmata*	WB	350-1100	10.2			Sikka et al. (1976)
4 Species of unicellular algae	WB	3200-7300	0.2	7 days	stat	Hollister et al. (1975)
Blue crab, 5-day larva (*Callinectes sapidus*)	WB	1100	0.1	3 weeks	stat[a]	Bookhout & Costlow (1975)
Blue crab, 15-day larva (*Callinectes sapidus*)	WB	3000	0.01		stat[a]	Bookhout & Costlow (1975)
Blue crab, megalopa (*Callinectes sapidus*)	WB	2000	0.01		stat[a]	Bookhout & Costlow (1975)

Table 8 (contd).

Organism	Organ	BCF	Concentration (μg/litre)	Exposure period	Conditions	Reference
Pink shrimp, larva (*Penaeus duorarum*)	WB	2600	0.1	3 weeks		Lowe et al. (1971)
	liver	24 000	0.1	3 weeks		
Amphipod (*Hyalla azteca*)	WB	2530	1	28 days	stat	de la Cruz & Naqvi (1973)
Fathead minnow	WB	51 400	0.37	56 days	flow	Huckins et al. (1982)
Chicken, adult male	fat	138	7.2 mg/kg diet	26 weeks		Medley et al. (1974)
	fat	103	7 μg/kg diet	20 weeks		
	fat	69	0.71 mg/kg diet	20 weeks		
Mallard	egg	2.4-2.8	1 and 10mg/kg diet	25 weeks		Hyde et al. (1973b)
Mallard, adult female	wings	3.6-5.5	1 and 10mg/kg diet	25 weeks		
	liver	1.5-3.8		25 weeks		
	fat	30		25 weeks		

[a] Static culture bowl method with a change to fresh medium and chemical each day.
WB - Whole body.
BCF - Bioconcentration factor: concentration in tissue/concentration in medium.

Advisory Committee, 1972).[a] Where point discharges of mirex have taken place, accumulation of residues in birds' eggs is indicative of widespread distribution of mirex in the local environment.

According to Naqvi & de la Cruz (1973a), habitat appeared to affect bioaccumulation, the highest residues being found in ponds (0.37 mg/litre), creeks (0.31 mg/litre), grassland (0.28 mg/kg), lakes (0.27 mg/litre), and estuaries (0.20 mg/litre). Within ecosystems, there appeared to be a hierarchy of accumulation. In an aquatic ecosystem, the following residue levels were found; annelids (0.63 mg/kg), crustaceans (0.44 mg/kg), insects (0.29 mg/kg), fish (0.26 mg/kg), and molluscs (0.15 mg/kg).

8.5 Population and Community Effects

Mirex residues in aquatic ecosystems do not appear to be directly toxic to algae and phytoplankton at environmentally realistic concentrations. However, mirex can be concentrated by various species of phytoplankton, which can thus serve as passive agents of transfer of mirex up the food chain. In addition, adsorption on organic material in sediments results in a high toxicant input to detritus feeders (Leffler, 1975). Bioconcentration in aquatic species (Table 8) is a very common problem. Transportation of mirex in a food chain was demonstrated when grass shrimp, which had been poisoned by being individually fed one particle of mirex bait, were fed to juvenile blue crab. These crabs then died from mirex poisoning within 14 days of eating 1 - 4 shrimps (Lowe et al., 1971). As expected, predatory animals contain higher residues than omnivores or herbivores (Collins et al., 1974).

Sensitivity of larval and juvenile crustaceans to mirex is very significant because their survival success determines the fate of entire populations (Bookhout & Costlow, 1976). However, no massive die-offs or declines in population have been reported for crustacea (Markin et al., 1972). Dose-dependent secondary effects observed may be particularly important at low concentrations. In fish, bacterial infection and growth inhibition are secondary effects of mirex poisoning (Van Valin et al., 1968), but there have been no detailed field surveys of the effects of mirex on fish populations (Task Force on Mirex, 1977). Of importance to the aquatic community is the depressive effect of mirex on lugworm activity, which will delay trapping of pollutants in sediments (Schoor & Newman, 1976).

---

[a] Report to US Environmental Protection Agency.

There is evidence of accumulation of mirex in aquatic and terrestrial food chains to harmful levels. After 6 applications of mirex bait at 1.4 kg/ha, high mirex levels were found in some species; turtle fat contained 24.8 mg mirex/kg, kingfishers, 1.9 mg/kg, coyote fat, 6 mg/kg, oppossum fat, 9.5 mg/kg, and racoon fat, 73.9 mg/kg (Hyde et al., 1973a).

In a model ecosystem with a terrestrial-aquatic interface, sorgum seedlings were treated with mirex at 1.1 kg/ha (Metcalf et al., 1973). Caterpillars fed on sorgum seedlings and their faeces contaminated the water which contained algae, snails, *Daphnia*, mosquito larvae, and fish. After 33 days, the ecological magnification value was 219 for fish and 1165 for snails.

An area of early old-field treated with mirex showed less vegetation biomass and lower species diversity than an untreated old-field ecosystem (Cassita & Kricher, 1973). Most terrestrial invertebrates in the USA contained less than 0.1 mg mirex/kg residues, but some species, particularly scavengers (that eat bait directly) and predators, contained as much as 30 mg/kg. Mirex can cause temporary population decline in insects. A 0.3% granular formulation of mirex applied at 0.20 kg/ha (recommended rate for the control of fire ant) caused a significant reduction in carabid, staphylid, and cricket numbers, though spider numbers were unaffected (Reagan et al., 1972). It was noted that application of twice this recommended rate did not eliminate fire ants. Mirex applied as corncob bait and sprayed at 4.2 - 42 kg/ha in a mixed hardwood forest caused a decline in centipede numbers but not in numbers of spiders, millipedes, beetles, and scorpions (Lee, 1974). Leaf decomposition was significantly accelerated. As would be expected, control measures against fire ants cause destruction of general ant fauna (Markin et al., 1974a). However, a decline in an insect population is not permanent (Mirex Advisory Committee, 1972;[a] Lee, 1974). There was no effect of mirex treatment on population size in 18 insect species in the year following application to one ecosystem (Wojcik et al., 1975).

Most mammals living in areas treated with mirex contain mirex residues. These reach a maximum 1 - 3 months after application and decline significantly during the next 12 months (Wojcik et al., 1975). No toxic effects of mirex on wildlife have been recorded in Canada (Canada, Department of National Health and Welfare, 1977).

---

[a] Report to US Environmental Protection Agency.

Very low mirex levels in falcon eggs collected in eastern and northern Canada in 1975 indicated that there was no apparent indigenous problem with mirex contamination, since falcons are near the top of the food pyramid in these terrestrial ecosystems (Task Force on Mirex, 1977).

8.6 Effects on the Abiotic Environment

Levels in sediment samples collected from Lake Ontario indicated that mirex continues to accumulate in harbour and offshore sediments, although decreasing amounts are being deposited in more recent lake sediments (Scrudato & del Prete, 1982). Mirex-contaminated sediments are accumulating in deeper water (100 m) of the lake at about 2.2 - 7.0 mm/year. It was suggested that it might be 200 - 600 years before mirex-contaminated sediments were covered by 'clean' sediments (Halfon, 1981). In addition, natural and anthropogenic mixing of contaminated sediments would provide a continuing source of mirex for lake organisms.

8.7 Appraisal

Mirex is one of the most environmentally stable of the organochlorine insecticides. There are 2 clearly established routes of contamination of the environment, the first from the manufacture and industrial use of mirex and the second from its agricultural use in control programmes for fire ants. Mirex bioaccumulates at all trophic levels and biomagnifies in food chains. It degrades slowly and its breakdown products are as toxic and stable as the parent compound. Mirex is strongly adsorbed on sediments and only poorly soluble in water. These characteristics combine with biotic factors, such as inhibition of activity of burrowing detritus feeders, to guarantee environmental accumulation of mirex and to slow down its removal from sediment and the covering of contaminated sediment layers with clean material.

A major pathway of mirex movement is from sediments or water into scavengers or herbivores. These are eaten by predatory invertebrates that are themselves ultimately eaten by vertebrates. This is the classic food-chain concentration of a contaminant. Biomagnification in the food chain is further encouraged by the delayed mortality typical of mirex poisoning. Because of its delayed effects, mirex shows a wide range of acute toxicity in different species. Chronic toxicity is a better indicator of the true toxicity of mirex and is uniformly high. This delayed effect appears to result from its high rate of uptake and slow rates of metabolism and excretion.

Effects on organisms combined with its persistence suggest that mirex presents a long-term hazard for the environment. Mirex induces pervasive chronic physiological and biochemical disorders in various vertebrates. Aquatic crustaceans show extreme sensitivity to the compound, and game birds and fish feeding close to manufacturing plants accumulate enough mirex to constitute a health hazard. Some birds feed in contaminated areas and then migrate to other areas, resulting in the unpredictable dispersal of mirex.

Although general environmental levels of mirex are low, it is widespread. The broadcast use of mirex in agriculture poses the greatest threat in increasing this contamination.

9. PREVIOUS EVALUATIONS OF MIREX BY INTERNATIONAL BODIES

IARC (1979) evaluated the carcinogenic hazard resulting from exposure to mirex and concluded that "there is sufficient evidence for its carcinogenicity to mice and rats. In the absence of adequate data in humans, it is reasonable, for practical purposes, to regard mirex as if it presented a carcinogenic risk to humans".

No acceptable daily intake (ADI) for mirex has been advised by FAO/WHO.

Over recent years, official registrations for a number of uses of mirex have been withdrawn in several countries for various reasons. Details can be obtained from IRPTC.

Regulatory standards established by national bodies in 12 different countries (Argentina, Brazil, Czechoslovakia, the Federal Republic of Germany, India, Japan, Kenya, Mexico, Sweden, the United Kingdom, the USA, and the USSR) and the EEC can be obtained from the IRPTC (International Register of Potentially Toxic Chemicals) Legal File (IRPTC, 1983).

10. EVALUATION OF HEALTH RISKS FOR MAN AND EFFECTS ON THE ENVIRONMENT

10.1 Mirex Toxicity

Mirex is moderately toxic in single-dose animal studies (oral LD_{50} values range from 365 - 3000 mg/kg body weight). It can enter the body via inhalation, ingestion, and via the skin.

It is one of the most stable pesticides in use today. It accumulates in adipose tissue and biomagnifies in food chains. Excretion is slow and elimination half-lives can extend over many months.

The most sensitive effects of repeated exposure in experimental animals are principally associated with the liver, and these have been observed with doses as low as 1.0 mg/kg diet (0.05 mg/kg body weight per day), the lowest dose tested.

At higher dose levels, it is fetotoxic (25 mg/kg in diet) and teratogenic (6.0 mg/kg per day).

Mirex was not generally active in short-term tests for genetic activity. There is sufficient evidence of its carcinogenicity in mice and rats.

No data on effects on human beings were available to the Task Group.

10.2 Exposure to Mirex

In the general population, food probably represents the major source of intake of mirex, fish, wild game, and meat being the main sources. Normally, such intake will be below established residue tolerances. Mirex may occur in breast milk but levels are very low or below detection limits.

No data are available, regarding occupational exposure.

10.3 Evaluation of Environmental Impact

Mirex is one of the most stable of the organochlorine insecticides. Although general environmental levels are low, it is widespread in the biotic and abiotic environment. Mirex is both accumulated and biomagnified. Mirex is strongly adsorbed on sediments and has a low water solubility.

The delayed onset of toxic effects and mortality is typical of mirex poisoning. The long-term toxicity of mirex is uniformly high. Mirex is toxic for a range of aquatic organisms, with crustacea being particularly sensitive. Mirex induces pervasive long-term physiological and biological disorders in vertebrates.

Although no field data are available, the adverse effects of long-term exposure to low levels of mirex combined with its persistence suggest that the use of mirex presents a long-term environmental risk.

10.4 Conclusions

1. No data on human health effects are available in connection with occupational exposure to mirex. Based on the findings in mice and rats, this chemical should be considered, for practical purposes, as being potentially carcinogenic for human beings.

2. For the same reason, reservations must remain about the safety of this chemical in food, despite the relatively low residues so far reported.

3. Effects on the organisms studied, as well as its persistence, suggest that mirex presents a long-term hazard for the environment.

4. Taking into account these considerations, it is felt that the use of this chemical for both agricultural and non-agricultural applications should be discouraged, except where there is no adequate alternative.

REFERENCES

ABSTON, P.A. & YARBROUGH, J.D. (1976) The *in vivo* effect of mirex on soluble hepatic enzymes in the rat. *Pestic. Biochem. Physiol.*, *6*: 192-197.

ALLEY, E.G., DOLLAR, D.A., LAYTON, B.R., & MINYARD, J.P., Jr (1973) Photochemistry of mirex. *J. agric. food Chem.*, *21*: 138-139.

ALLEY, E.G., LAYTON, B.R., & MINYARD, J.P., Jr (1974) Identification of the photoproducts of the insecticides mirex and kepone. *J. agric. food Chem.*, *22*: 442-445.

ANDRADE, P.S.L. & WHEELER, W.B. (1974a) Biodegradation of Mirex by sewage sludge organisms. *Bull. environ. Contam. Toxicol.*, *11*: 415-416.

ANDRADE, P.S.L. & WHEELER, W.B. (1974b) Mirex transformation products in the environment. *Abstracts of papers, 168th National Meeting of the American Chemical Society, Atlantic City, New Jersey, 8-13, September*, Baltimore, Maryland, Port City Press, 31 pp.

ANDRADE, P.S.L., WHEELER, W.B., & CARLSON, D.A. (1975) Identification of a mirex metabolite. *Bull. environ. Contam. Toxicol.*, *14*: 473-479.

ATALLAH, Y.H. & DOROUGH, H.W. (1975) Insecticide residues in cigarette smoke. Transfer and fate in rats. *J. agric. food Chem.*, *23*: 64-71.

ATKINS, E.L., KELLUM, D., & NEWMAN, K.J. (1975) *Toxicity of pesticides to honey bees*, California, University of California, Division of Agriculture Sciences, 4 pp (Leaflet 2286).

BAETCKE, K.P., CAIN, J.D., & POE, W.E. (1972) Residues in fish, wildlife and estuaries: mirex and DDT residues in wildlife and miscellaneous samples in Mississippi - 1970. *Pestic. Monit. J.*, *6*: 14-22.

BAKER, M.F. (1963) *Field-pen tests of mirex with bobwhite quail*, Washington DC, US Department of the Interior, Fish and Wildlife Service, PP. 116, 117, 124 (Circular No. 199).

BAKER, R.C., COONS, L.B., MAILMAN, R.B., & HODGSON, E. (1972) Induction of hepatic mixed-function oxidases by the insecticide mirex. Environ. Res., 5: 418-424.

BAKER, R.D. & APPLEGATE, H.G. (1974) Effect of ultraviolet radiation on the persistence of pesticides. Tex. J. Sci., 25: 53-59.

BELL, M.A., EWING, R.A., & LUTZ, G.A. (1978) Reviews of the environmental effects of pollutants. 1. mirex and kepone, Washington DC, US Environmental Protection Agency (US EPA Report No. EPA-600/1-78-01, US NTIS PB 80-12595).

BEVENUE, A., OGATA, J.N., TENGAN, L.S., & HYLIN, J.W. (1975) Mirex residues in wildlife and soils. Hawaiian pineapple-growing areas, 1972-74. Pestic. Monit. J., 9: 141-149.

BLUS, L.J., PATTEE, O.H., HENRY, C.J., & PROUTY, R.M. (1983) First records of chlordane-related mortality in wild birds. J. Wildl. Manage., 47: 196-198.

BOND, C.A., WOODHAM, D.W., AHRENS, E.H., & MEDLEY, J.G. (1975) The cumulation and disappearance of mirex residues. Part 2. In milk and tissues of cows fed 2 concentrations of the insecticide in their diet. Bull. environ. Contam. Toxicol., 14: 25-31.

BONG, R.L. (1975) Determination of hexachlorobenzene and mirex in fatty products. J. Assoc. Off. Anal. Chem., 58: 557-561.

BONG, R.L. (1977) Collaborative study of the recovery of hexachlorobenzene and mirex in butterfat and fish. J. Assoc. Off. Anal. Chem., 60: 229-232.

BOOKHOUT, C.G. & COSTLOW, J.D. (1975) Effects of mirex on larval development of blue crab. Water air soil Pollut., 4: 113.

BOOKHOUT, C.G. & COSTLOW, J.D. (1976) Effects of mirex, methoxychlor and melathion on development of crabs, Beaufort, North Carolina, Duke University, 96 pp (US EPA 600/3-76-007).

BOOKHOUT, C.G., WILSON, A.J., DUKE, T.W., & LOWE, J.I. (1972) Effects of mirex on the larval development of two crabs. Water air soil Pollut., 1: 165-180.

BORTHWICK, P.W., DUKE, T.W., WILSON, A.J., Jr, LOWE, J.I., PATRICK, J.M., Jr, & OBERHEU, J.C. (1973) Residues in fish, wildlife, and estuaries: accumulation and movement of mirex in selected estuaries of South Carolina, 1969-71. Pestic. Monit. J., 7(1): 6-26.

BORTHWICK, P.W., COOK, G.H., & PATRICK, J.M., Jr (1974) Residues in fish, wildlife, and estuaries: mirex residues in selected estuaries of South Carolina-June 1972. Pestic. Monit. J., 7(3/4): 144-145.

BROWN, L.R., ALLEY, E.G., & COOK, D.W. (1975) The effect of mirex and carbofuran on estuarine microorganisms, Cornwallis, Oregon, US Environmental Protection Agency, Office of Research & Development, Natural Environmental Research Center, 47 pp (US EPA 660/3-75-024).

BUTLER, P.A. (1963) A review of fish and wildlife service investigations during 1961 and 1962. In: George, J.L., ed. Commercial fisheries investigations, pesticide - wildlife series, Washington DC, US Department of the Interior, Fish and Wildlife Services, pp. 11-25 (Circular No. 167).

BYARD, J.L., KOEPKE, U.C., ABRAHAM, R., GOLDBERG, L., & COULSTON, F. (1975) Biochemical changes in the liver of mice fed mirex. Toxicol. appl. Pharmacol., 33: 70-77.

CABRAL, J.R.P., RAITANO, F., MOLLNER, T., BRONCZYK, S., & SHUBIK, P. (1979) Acute toxicity of pesticides in hamsters. Toxicol. appl. Pharmacol., 48: A192.

CANADA, DEPARTMENT OF NATIONAL HEALTH AND WELFARE (1977) Mirex, environmental health criteria document (Report 77-EHD-12).

CARLSON, D.A., KONYHA, K.D., WHEELER, W.B., MARSHALL, G.P., & ZAYLSKIE, R.G. (1976) Mirex in the environment: its degradation to kepone and related compounds. Science, 194: 939-941.

CASSISTA, A.J. & KRICHER, J.C. (1973) Possible phytotoxic effects of the insecticide mirex in an early old field ecosystem. Plant Physiol., 51: 22 (Abstract).

CHAMBERS, J.E. & TREVATHAN, C.A. (1983) Effect of mirex, dechlorinated mirex derivatives and chlordecone on microsomal mixed-function oxidase activity and other hepatic parameters. Toxicol. Lett., 16: 109-115.

CHERNOFF, N., SCOTTI, T.M., & LINDER, R.E. (1976) Cataractogenic properties of mirex in rats and mice with notes on kepone. Toxicol. appl. Pharmacol., 37: 188.

CHU, I., VILLENEUVE, D.C., SECOURS, V., RUDDICK, J., BECKING, G.C., & VALLI, V.E. (1980a) 2,8-Dihydromirex: a twenty-eight day subacute study in the rat. J. environ. Sci. Health, B15: 87-107.

CHU, I., VILLENEUVE, D.C., VALLI, V.E., & REYNOLDS, L.M. (1980b) Short-term study of the combined effects of mirex, photomirex and kepone to halogenated biphenyls in the rat. J. Toxicol. environ. Health, 6: 421-432.

CHU, I., VILLENEUVE, D.C., SECOURS, V., VALLI, V.E., & BECKING, G.C. (1981) Effects of photomirex and mirex on reproduction in the rat. Toxicol. appl. Pharmacol., 60(3): 549-556.

CHU, I., VILLENEUVE, D.C., & VIAN, A. (1982) Tissue distribution of elimination of photomirex in squirrel monkeys. Bull. environ. Contam. Toxicol., 29: 434-439.

COLLINS, H.L., DAVIS, J.R., & MARKIN, G.P. (1973) Residues of mirex in channel catfish and other aquatic organisms. Bull. environ. Contam. Toxicol., 10: 73-77.

COLLINS, H.L., MARKIN, G.P., & DAVIS, J.R. (1974) Residue accumulation in selected vertebrates following a single aerial application of mirex bait, Louisiana-1971-72. Pestic. Monit. J., 8: 125-130.

COOLEY, N.R., KELTNER, J.M., & FORRESTER, J. (1972) Mirex and aroclor 1254: effect on an accumulation by Tetrahymena pyriformis strain W. J. Protozool., 19: 636-638.

COPE, O.B. (1966) Contamination of the freshwater ecosystem by pesticides. J. appl. Ecol. Suppl., 3: 33-44.

DAVISON, K.L. & COX, J.H. (1974) Some effects of mirex on chickens, quail and rats. Fed. Proc., 33: 220.

DAVISON, K.L., COX, J.H., & GRAHAM, C.K. (1975) The effect of mirex on the reproduction of Japanese quail and on the characteristics of eggs from Japanese quail and chickens. Arch. environ. Contam. Toxicol., 3: 84-95.

DAVISON, K.L., MOLLENHAUER, H.H., YOUNGER, R.L., & COX, J.H. (1976) Mirex-induced hepatic changes in chickens, Japanese quail and rats. Arch. environ. Contam. Toxicol., 4: 469-482.

DE LA CRUZ, A.A. & NAQVI, S.M. (1973) Mirex incorporation in the environment: uptake in aquatic organisms and effects on the rates of photosynthesis and respiration. Arch. environ. Contam. Toxicol., 1: 255-264.

DE LA CRUZ, A.A. & RAJANNA, B. (1975) Mirex incorporation in the environment: uptake and distribution in crop seedlings. Bull. environ. Contam. Toxicol., 14: 38-42.

DIETZ, D.D. & MCMILLAN, D.E. (1978) Effects of mirex and kepone on schedule controlled responding. Pharmacologist, 20: 225.

DILLING, W.L. & DILLING, M.L. (1967) Pentacyclodecane chemistry. III. Fragmentation patterns of pentacyclodecane derivatives on electron impact. Tetrahedron, 23: 1225-1233.

DRIVER, D., BREWER, R.N., & COTTIER, G.J. (1976) Pesticide residues in eggs and chicks from laying hens fed low levels of chlorinated hydrocarbon pesticides. Poult. Sci., 55: 1544-1549.

FABACHER, D.L. & HODGSON, E. (1976) Induction of hepatic mixed-function oxidase enzymes in adult and neonatal mice by kepone and mirex. Toxicol. appl. Pharmacol., 38: 71-77.

FORD, J.H., HAWTHORNE, J.C., & MARKIN, G.P. (1973) Residues of mirex and other chlorinated hydrocarbon insecticides in beef fat-1971. Pestic. Monit. J., 7: 87-94.

FULFS, J., ABRAHAM, R., DROBECK, B., PITTMAN, K., & COULSTON, F. (1977) Species differences in the hepatic response to mirex: ultrastructural and histochemical studies. Ecotoxicol. environ. Saf., 1: 327-342.

FULLER, G.B. & DRAPER, S.W. (1975) Effect of mirex on induced ovulation in immature rats. Proc. Soc. Exp. Biol. Med., 148: 414-417.

GAINES, T.B. (1969) Acute toxicity of pesticides. Toxicol. appl. Pharmacol., 14: 5-534.

GAINES, T.B. & KIMBROUGH, R.D. (1970) Oral toxicity of mirex in adult and suckling rats. Arch. environ. Health, 21: 7-14.

GIBSON, J.R., IVIE, G.W., & DOROUGH, H.W. (1972) Fate of mirex and its major photodecomposition product in rats. J. agric. food Chem., 20: 1246-1248.

GLICK, B. (1974) Antibody-mediated immunity in the presence of mirex and DDT. Poult. Sci., 53: 1476-1485.

GRABOWSKI, C.T. & PAYNE, D.B. (1983) The causes of perinatal death induced by prenatal exposure of rats to the pesticide mirex. Part. 1. Pre-parturition observations of the cardiovascular system. Teratology, 27: 7-11.

GUECKEL, W. (1973) A method for determining the volatility of active ingredients used in plant protection. Pestic. Sci., 4: 137-147.

GUNBY, P. & PRESTON, M. (1979) Fire ants are stinging nine southern states. J. Am. Med. Assoc., 241: 2689-2690.

HALFON, E. (1981) Error analysis and stimulation behaviour in Lake Ontario, Burlington, Ontario, National Water Research Institute, Canada Centre for Inland Waters.

HALLETT, D.J., NORSTROM.F.J., ONUSUKA, F.I., COMBA, M.E., & SAMPSON, R. (1976) Mass spectral confirmation and analysis by the Hall detector of mirex and photomirex in herring gulls from Lake Ontario. J. agric. food Chem., 24: 1189-1193.

HALLETT, D.J., KHERA, K.S., STOLTZ, D.R., CHU, I., VILLENEUVE, D.C., & TRIVETT, G. (1978) Photomirex: synthesis and assessment of acute toxicity, tissue distribution and mutagenicity. J. agric. food Chem., 26: 388-391.

HARTMANN, H.C. (1971) Gas chromatography detectors. Anal. Chem., 43: 113A-125A.

HEATH, R.G. & SPANN, J.W. (1973) Reproduction and related residues in birds fed mirex. Ind. Med., 42: 22.

HEATH, R.G., SPANN, J.W., HILL, E.F., & KRIETZER, J.F. (1972) Comparative dietary toxicities of pesticides to birds, Washington DC, US Department of the Interior, US Fish and Wildlife Services, 57 pp (Special Scientific Wildlife Report No. 152).

HILL, E.F., HEATH, R.G., SPANN, J.W., & WILLIAMS, J.D. (1975) Lethal dietary toxicities of environmental pollutants to birds, Washington DC, US Department of the Interior, US

Fish and Wildlife Services, 61 pp (Special Scientific Wildlife Report No. 191).

HOLLISTER, T.A., WALSH, G.E., & FORESTER, J. (1975) Mirex and marine unicellular algae: accumulation, population growth and oxygen evolution. Bull. environ. Contam. Toxicol., 14: 753-759.

HORWITZ, W. (1975) Official methods of analysis of the Association of Official Analytical Chemists, 12th ed., Washington DC, AOAC.

HUCKINS, J.N., STALLING, D.L., PETTY, J.D., BUCKLER, D.R., & JOHNSON, B.T. (1982) Fate of kepone and mirex in the aquatic environment. J. agric. food Chem., 30: 1020-1027.

HYDE, K.M. (1973) Studies of the responses of selected wildlife to mirex bait exposure. Diss. Abstr. Int., 33: 3693.

HYDE, K.M., GRAVES, J.B., FOWLER, J.F., BONNER, F.L., IMPSON, J.W., NEWSOM, J.D., & HAYBOOD, J. (1973a) Accumulation of mirex in food chains. Louisiana Agric., 17: 10-11.

HYDE, K.M., GRAVES, J.B., WATTS, A.B., & BONNER, F.L. (1973b) Reproductive success of mallard ducks fed mirex. J. Wildl. Manage., 37: 479-484.

HYDE, K.M., STOKES, S., FOWLER, J.F., GRAVES, J.B., & BONNER, F.L. (1974) The effect of mirex on channel catfish production. Trans. Am. Fish. Soc., 103: 366-369.

IARC (1979) Some halogenated hydrocarbons, Lyons, International Agency for Research on Cancer, (Monographs on the Evaluation of the Carcinogenic Risk of Chemicals to Humans, Vol. 20).

INNES, J.R.M., ULLAND, B.M., VALERIO, M.G., PETRUCELLI, L., FISHBEIN, L., MART, O.R., PALLOTTA, A.J., BATES, R.R., FALK, R., GART, J.J., KLEIN, M., MITCHELL, I., & PETERS, J. (1969) Bioassay of assay of pesticides and industrial chemicals for tumorigenicity in mice: a preliminary note. J. Natl Cancer Inst., 42: 101-1114.

IRPTC (1983) Legal file, Vols. 1 & 2, Geneva, International Register of Potentially Toxic Chemicals, United Nations Environment Programme.

IVERSON, F. (1976) Induction of paraoxon dealkylation by hexachlorobenzene (HCB) and mirex. J. agric. food Chem., 24: 1238-41.

IVIE, G.W., DOROUGH, H.W., & ALLEY, E.G. (1974a) Photodecomposition of mirex on silica gel chromatoplates exposed to natural and artificial light. J. agric. food Chem., 22: 933-935.

IVIE, G.W., GIBSON, J.R., BRYANT, H.E., BEGIN, J.J., BARNETT, J.R., & DOROUGH, H.W. (1974b) Accumulation, distribution and excretion of mirex-ssM, ^{14}C in animals exposed for long periods to the insecticide in the diet. J. agric. food Chem., 22: 646-653.

JOHNSON, W.W. & FINLEY, M.T. (1980) Handbook of acute toxicity of chemicals to fish and aquatic invertebrates, Washington DC, US Department of the Interior, US Fish and Wildlife Services, 56 pp (Resource Publication No. 137).

JOHNSON, W.W. & MAYER, F.L. (1973) Pesticides and the aquatic environment. Bull. Mo. Acad. Sci. Suppl., 1: 21-37.

JONES, A.S. & HODGES, C.S. (1974) Persistence of mirex and its effects on soil microorganisms. J. agric. food Chem., 22: 435-439.

KAISER, K.L.E. (1978) The rise and fall of mirex. Environ. Sci. Technol., 12: 520-528.

KAMINSKY, L.S., PIPER, L.J., MCMARTIN, D.M., & FASCO, M.J. (1978) Induction of hepatic microsomal cytochrome P-450 by mirex and kepone. Toxicol. appl. Pharmacol., 43: 327-338.

KENDALL, M.W. (1974) Acute hepatotoxic effects of mirex in the rat. Bull. environ. Contam. Toxicol., 12: 617-621.

KENDALL, M.W. (1979) Light and electron microscopic observations of the acute, sublethal hepatotoxic effects of mirex in the rat. Arch. environ. Contam. Toxicol., 8: 25-41.

KENDALL, R.J., NOBLET, R., SENN, L.H., & HOLMAN, J.R. (1978) Toxicological studies with mirex in bobwhite quail. Poult. Sci., 57: 1539-45.

KENNEDY, M.V., HOLLOMAN, M.E., & HUTTO, F.Y. (1977) Thermal degradation of selected fungicides and insecticides. In: A Symposium at the 174th Meeting of ACS, Chicago, August 29-September 2, 1977; "Disposal and Decontamination of

Pesticides", pp. 81-99 (American Chemical Society Symposium Series, 1973).

KHERA, K.S., VILLENEUVE, D.C., TERRY, G., PANOPIO, L., NASH, L., & TRIVETT, G. (1976) Mirex: a teratogenicity, dominant lethal and tissue distribution study in rats. Food cosmet. Toxicol., 14: 25-29.

KNOEVENAGEL, K. & HIMMELREICH, R. (1976) Degradation of compounds containing carbon atoms by photooxidation in the presence of water. Arch. environ. Contam. Toxicol., 4: 324-333.

KOBYLINSKI, G.J. & LIVINGSTON, R.J. (1975) Movement of Mirex from sediment and uptake by the Hogchoker, Trinectes maculatus. Bull. environ. Contam. Toxicol., 14: 692-698.

KRITCHER, Y.C., UREY, J.C., & HAWES, M.L. (1975) The effects of mirex and methoxychlor on the growth and productivity of Chlorella pyrenoidosa. Bull. environ. Contam. Toxicol., 14: 617-620.

KUTZ, F.W., YOBS, A.R., JOHNSON, W.G., & WIERSMA, G.B. (1974) Mirex residues in human adipose tissue. Environ. Entomol., 3: 882-885.

LANE, H. (1973) Influence of food processing on mirex. Diss. Abstr. Int., 34: 2685B.

LANE, R.H., GRODNER, R.M., & GRAVES, J.L. (1976) Irradiation studies of mallard duck eggs material containing mirex. J. agric. food Chem., 24: 192-193.

LARSON, P.S., EGLE, J.L., Jr, HENNIGAR, G.R., & BORZELLECA, J.F. (1979) Acute and subchronic toxicity of mirex in the rat, dog, and rabbit. Toxicol. appl. Pharmacol., 49: 271-277.

LEE, B.J. (1974) Effects of mirex on litter organisms and leaf decomposition in a mixed hardwood forest in Athens, Georgia. J. environ. Qual., 3: 305-311.

LEE, J.H., SYLVESTER, J.R., & NASH, C.E. (1975) Effects of mirex and methoxychlor on juvenile and adult striped mullet, Mugil cephalus. Bull. environ. Contam. Toxicol., 14: 180-186.

LEFFLER, C.W. (1975) Effects of ingested mirex and DDT on juvenile Callinectes sapidus (Rathburn). Environ. Pollut., 8: 283-300.

LEWIS, R.G., HANISCH, R.C., MACLEOD, K.E., & SOVOCOL, G.W. (1976) Photochemical confirmation of mirex in the presence of polychlorinated biphenyls. J. agric. food Chem., 24: 1030-1035.

LEWIS, R.G., BROWN, A.R., & JACKSON, M.D. (1977) Evaluation of polyurethane biphenyls and polychlorinated naphthalenes in ambient air. Anal. Chem., 49: 1668-1672.

LLOYD, F.A., CAIN, C.E., MAST, J., CRITTENDEN, J., & THIEDE, H. (1974) Results of pesticide analysis of human maternal blood. J. Miss. Acad. Sci., 19: 79-84.

LOWE, J.I., WILSON, P.D., & DAVISON, R.B. (1970) Effects of mirex on crabs, shrimps, and fish. In: Progress Report of the Bureau of Commercial Fisheries Center for Estuarine and Menhaden Research Pesticide Field Station, Gulf Breeze, Florida, Fiscal Year 1969, Washington DC, US Department of the Interior, pp. 22-23 (Circular No. 335).

LOWE, J.I., PARRISH, P.R., WILSON, A.J., WILSON, P.D., & DUKE, T.W. (1971) Effects of mirex on selected estuarine organisms. In: Trefethen, J.B., ed. Transactions of the 36th American Wildlife and Natural Resources Conference, Washington DC, Wildlife Institute, Vol. 36, pp. 171-186.

LUDKE, J.L., FINLEY, M.T., & LUSK, C. (1971) Toxicity of mirex to crayfish, Procambarus blandingi. Bull. environ. Contam. Toxicol., 6: 89-95.

LUNZ, J.D. (1978) Habitat development field investigations Wildmill Point marsh development site James River, Virginia Appendix E. Environmental impact of marsh development with dredged material: metals and chlorinated hydrocarbon compounds in marsh soils 6 vascular plant tissues (Report ISS WES-TR-D-77-23-APP-E No. AD-A062).

MCARTHUR, M.L.B., FOX, G.A., PEAKALL, D.B., & PHILOGENE, B.J.R. (1983) Ecological significance of behavioural and hormonal abnormalities in breeding ring doves fed on organochlorine chemical mixture. Arch. environ. Contam. Toxicol., 12: 343-353

MACFARLANE, J., DIRKS, T., & UK, S. (1975) Symptoms of mirex, dieldrin and DDT poisoning in the field cricket Gryllus pensylvanicus, and effect on activity of the central nerve cord. Pestic. Biochem. Physiol., 5: 57-64.

MCKENZIE, M.D. (1970) Fluctuations in abundance of the blue crab and factors affecting mortalities, South Carolina, South

Carolina Wildlife Resources Department, Marine Resources Division, 45 pp (Technical Report No. 1).

MADHUKAR, B.V. & MATSUMURA, F. (1979) Comparison of induction patterns of rat hepatic microsomal mixed-function oxidases by pesticides and related chemicals. Pestic. Biochem. Physiol., 11: 301-308.

MARKIN, G.P., FORD, J.H., & HAWTHORNE, J.C. (1972) Mirex residue in wild populations of edible red crayfish, Porcambarus clarkii. Bull. environ. Contam. Toxicol., 8: 369-374.

MARKIN, G.P., O'NEAL, J., & COLLINS, H.L. (1974a) Effects of mirex on the general ant fauna of a treated area in Louisana. Environ. Entomol., 3: 895-898.

MARKIN, G.P., COLLING, H.L., & SPENCE, J.H. (1974b) Residues of the insecticide mirex following aerial treatment of Cat Island. Bull. environ. Contam. Toxicol., 12: 233-240.

MARKIN, G.P., COLLINS, H.L., & ONEAL, J. (1975) Control of imported fire ants with winter applications of micro-encapsulated mirex bait. J. Econ. Ent., 68: 711.

MATSUMURA, F. (1975) Toxicology of insecticides, New York, Plenum Press.

MEDLEY, J.G., BOND, C.A., & WOODHAM, D.W. (1974) The cumulation and disappearance of mirex residues. I. In tissues of roosters fed four concentrations of mirex in their feed. Bull. environ. Contam. Toxicol., 11: 217-223.

MEHENDALE, H.M. (1976) Mirex-induced suppression of biliary excretion of polychlorinated biphenyl compounds. Toxicol. appl. Pharmacol., 36: 369-381.

MEHENDALE, H.M. (1977a) Mirex-induced impairment of hepatobiliary function. Suppressed biliary excretion of imipramine and sulfobromophthalein. Drug Metab. Dispos., 5: 56-62.

MEHENDALE, H.M. (1977b) Chemical activity, absorption, retention, metabolism and elimination of hexachlorocyclo-pentadiene. Environ. Health Perspect., 21: 275-278.

MEHENDALE, H.M. (1979) Modification of hepatobiliary function by toxic chemicals. Fed. Proc. Fed. Am. Soc. Exp. Biol., 38: 2240-2245.

MEHENDALE, H.M., FISHBEIN, L., FIELDS, M., & MATTHEWS, H.B. (1972) Fate of ^{14}C-mirex in the rat and plants. Bull. environ. Contam. Toxicol., 8: 200-207.

MEHENDALE, H.M., CHEN, P.R., FISHBEIN, L., & MATTHEWS, H.B. (1973) Effect of mirex on the activities of various rat hepatic mixed-function oxidases. Arch. environ. Contam. Toxicol., 1(3): 245-254.

MEHENDALE, H.M., HO, I.K., & DESAIAH, D. (1979) Possible molecular mechanism of mirex-induced hepatobiliary dysfunction. Drug Metab. Dispos., 7: 28-33.

MES, J. & DAVIES, D.J. (1978) Variation in the polychlorinated biphenyl and organochlor pesticide residues during human breastfeeding and its diurnal pattern. Chemosphere, 7: 699- 706.

MES, J., DAVIES, D.J., & MILES, W. (1978) Traces of mirex in some Canadian human milk samples. Bull. environ. Contam. Toxicol., 19: 564-570.

METCALF, R.L., KAPOOR, I.P., LU, P-Y., SCHUTH, C.K., & SHERMAN, P. (1973) Model ecosystem studies of the environmental fate of six organochlorine pesticides. Environ. Health Perspect., 4: 35-44.

MUNCY, R.J. & OLIVER, A.D. (1963) Toxicity of ten insecticides to the red crayfish, Procambarus clarkii (Giraud). Trans. Am. Fish. Soc., 92: 428-431.

NABER, E.C. & WARE, G.W. (1965) Effect of kepone and mirex on reproductive performance in the laying hen. Poult. Sci., 44: 875-880.

NAQVI, S.M. & DE LA CRUZ, A.A. (1973a) Mirex incorporation in the environment. Residues in non-target organisms - 1972. Pestic. Monit. J., 7: 104-111.

NAQVI, S.M. & DE LA CRUZ, A.A. (1973b) Mirex incorporation in the environment: toxicity in selected freshwater organisms. Bull. environ. Contam. Toxicol., 10: 305-308.

O'NEAL, J., MARKIN, G.P., & COLLINS, H.L. (1974) Effects of mirex on general ant fauna of a treated area in Louisiana. Environ. Entomol., 3: 895-898.

PEELER, D.F. (1976) Open field activity as a function of pre-weaning or generational exposure to mirex. J. Miss. Acad. Sci., 21: 58.

PESTICIDE CHEMICAL NEWS (1976) June 9th, p. 8.

PITTMAN, K.A., WIENER, M., & TREBLE, D.H. (1976) Mirex kinetics in the rhesus monkey. II. Pharmacokinetic model. Drug Metab. Dispos., 4: 288-295.

PITZ, E.P., ROURKE, D., ABRAHAM, R., & COULSTON, F. (1979) Alterations in hepatic microsomal proteins of mice administered mirex orally. Bull. environ. Contam. Toxicol., 21: 344-351.

PLAPP, F.W. (1973) Mirex: toxicity, tolerance and metabolism in the house fly Musca domestic L. Environ. Entomol., 2: 1058-1061.

PRITCHARD, J.B., GUARINO, A.M., & KINTER, W.B. (1973) Distribution, metabolism and excretion of DDT and mirex by a marine Teleost, the winter flounder. Environ. Health Perspect., 4:45-54.

RAJANNA, B. & DE LA CRUZ, A.A. (1975) Mirex incorporation in the environment. Phytotoxicity on germination, emergence and early growth of seedlings. Bull. environ. Contam. Toxicol., 14: 77-82.

RAO, D.S.V.S. & GLICK, B. (1977) Pesticide effects on the immune response and metabolic activity of chicken lymphocytes. Proc. Soc. Exp. Biol. Med., 154:27-29.

REAGAN, T.E., COBURN, G., & HENSLEY, S.D. (1972) Effects of mirex on the arthropod fauna of a Louisiana sugar cane field. Environ. Entomol., 1: 588-591.

REITER, L. (1977) Behavioral toxicology: effects of early postnatal exposure to neurotoxins on development of locomotor activity in the rat. J. occup. Med., 19: 201-204.

REITER, L., KIDD, K., LEDBETTER, G., CHERNOFF, N., & GRAY, L.E., Jr (1977) Comparative behavioral toxicology of mirex and kepone in the rat. Toxicol. appl. Pharmacol., 41: 143.

ROBINSON, K.M. & YARBROUGH, J.D. (1968) Liver response to oral administration of mirex in rats. Pestic. Biochem. Physiol., 8: 65-72.

ROBINSON, K.M. & YARBROUGH, J.D. (1978) A study of liver function in rats with mirex-induced enlarged livers. Pestic. Biochem. Physiol., 9: 61-64.

SANDHU, S.S., WARREN, W.J., & NELSON, P. (1978) Pesticidal residue in rural potable water. J. Am. Water Works Assoc., 70: 41-45.

SAVAGE, E.P. (1976) National study to determine levels of chlorinated hydrocarbon insecticides in human milk: 1975-76, Fort Collins, Colorado, Colorado State University, Epidemiologic Studies Center (Report, Iss. EPA/540/9-78/005, order no. PB284393).

SCHOOR, W.P. (1974) Accumulation of mirex in the adult blue crab, Callinectes sapidus. Bull. environ. Contam. Toxicol., 12: 136-137.

SCHOOR, W.P. & NEWMAN, S.M. (1976) The effect of mirex on the burrowing activity of the lugworm Arenicola cristala. Trans. Am. Fish. Soc., 105: 700-703.

SCHRAUZER, G.N. & KATZ, R.N. (1978) Reductive dechlorination and degradation of mirex and kepone with vitamin B12. Bioinorg. Chem., 9: 123-143.

SCRUDATO, R.J. & DEL PRETE, A. (1982) Lake Ontario sediment - mirex relationships. J. Great Lake Res., 8: 659-699.

SHACKELFORD, W.M. & KEITH, L.H. (1976) Frequency of organic compounds, Athens, Georgia, US Environmental Protection Agency, p. 170 (US EPA 600/4-76-062).

SIKKA, H.C., BUTLER, G.L., & RICE, C.P. (1976) Effects, uptake and metabolism of methoxychlor, mirex and 2,4-D in seaweeds, Gulf Breeze, Florida, US Environmental Protection Agency Research Laboratory, NTIS, 48 pp (EPA 600/3-76-048).

SINGH, A., VILLENEUVE, D.C., BHATNAGAR, M.K., & VALLI, V.E.O. (1982) Ultrastructure of the thyroid glands of rats fed photomirex: an 18-month recovery study. Toxicology, 23: 309-319.

SMILLIE, NICHOLSON, MINESY, DUNOLKE, & REES (1977) Organics in drinking-water. Part II, Ontario, Ministry of the Environment (OTC, 7703).

SMREK, A.L., ADAMS, S.R., LIDDLE, J.A., & KIMBROUGH, R.D. (1977) Pharmacokinetics of mirex in goats. 1. Effect on reproduction and lactation. J. agric. food Chem., 25: 945-947.

SMREK, A.L., ADAMS, S.R., LIDDLE, J.A., & KIMBROUGH, R.D. (1978) Pharmacokinetics of mirex in goats. 2. Residue tissue levels, transplacental passage during recovery. J. agric. food Chem., 26: 945-947.

SPENCE, J.H. & MARKIN, G.P. (1974) Mirex residues in the physical environment following a single bait application, 1971-72. Pestic. Monit. J., 8: 135-139.

STEIN, V.B., PITTMAN, K.A., & KENNEDY, M.W. (1976) Characterization of a mirex metabolite from monkeys. Bull. environ. Contam. Toxicol., 15: 0-146.

STICKEL, W.H., GALYEN, J.A., DYRLAND, R.A., & HUGHES, D.L. (1973) Toxicity and persistance of mirex in birds. In: Proceedings of the 8th International American Conference on Toxicology and Occupational Medicine (Pesticide Symposium), Miami, Florida, pp. 437-467.

SUNDARAM, A., VILLENEUVE, D.C., CHU, I., SECOURS, V., & BECKING, G.C. (1980) Sub-chronic toxicity of photomirex in the female rat - results of 28-day and 90-day feeding studies. Drug chem. Toxicol., 3: 105-134.

SUTA, B.E. (1978) Human population exposures to mirex and kepone, Washington DC, US Environmental Protection Agency, p. 430 (US NTIS PB Report PB-285).

TAGATZ, M.E., BORTHWICK, P.W., & FORESTER, J. (1975) Seasonal effects of leached mirex on selected estuarine animals. Arch. environ. Contam. Toxicol., 3: 371-383.

TASK FORCE ON MIREX (1977) Mirex in Canada, Ottawa (Report to the Joint Department of Environment and National Health and Welfare Committee on Environmental Contaminants - Technical Report No. 77-1).

TEN NOEVER DE BRAUW, M.C., VAN INGEN, C., & KOEMAN, J.H. (1973) Mirex in seals. Sci. total Environ., 2: 196-198.

THORNE, B.M., TAYLOR, E., & WALLACE, T. (1978) Mirex and behavior in the Long-Evans rat. Bull. environ. Contam. Toxicol., 19: 351-359.

TUCKER, R.K. & CRABTREE, D.G. (1970) Handbook of toxicity of pesticides to wildlife, Washington DC, US Department of the Interior, Fisheries and Wildlife Service, 131 pp (Resource Publication No. 84).

ULLAND, B.M, PAGE, N.P., SQUIRE, R.A., WEISBURGER, E.K., & CYPHER, R.L. (1977) A carcinogenicity assay of mirex in Charles River rats. J. Natl Cancer Inst., 58: 133-140.

US EPA (1972) Effects of pesticides in water - a report to the United States, Washington DC, US Environmental Protection Agency, 145 pp (US EPA Report No. 16).

US EPA (1978) Kepone-mirex-hexachlorocyclopentadiene: an environmental assessment, Washington DC, US Environmental Protection Agency, pp. 36-50.

US NRC (1978) Kepone/mirex/hexachlorocyclopentadiene - an environmental assessment, Washington DC, US Department of Commerce (NTIS, PB 280-289).

US NTIS (1968) Evaluation of chronic, teratogenic, and mutagenic activities of selected pesticides and industrial chemicals, Vol. 1. Carcinogenic study, Washington DC, US Department of Commerce.

VAN VALIN, C.C., ANDREWS, A.K., & ELLER, L.L. (1968) Some effects of mirex on two warm-water fishes. Trans. Am. Fish. Soc., 97: 185-196.

VILLENEUVE, D.C., YAGMINAS, A.P., MARINO, I.A., CHU, I., & REYNOLDS, L.M. (1977) Effects of food deprivation in rats previously exposed to mirex, PESTAB/78/0118. Bull. environ. Contam. Toxicol., 18: 278-284.

VILLENEUVE, D.C., KHERA, K.S., TRIVETT, G., FELSKY, G., NORSTROM, R.J., & CHU, I. (1979a) Photomirex: a teratogenicity and tissue distribution study in the rabbit. J. Environ. Sci. Health, Part B, Pestic. Food Contam. Agric. Wastes, 14: 171-180.

VILLENEUVE, D.C., RITTER, L., FELSKY, G., NORSTROM, R., MARINO, I.A., VALLI, V.E., CHU, I., & BECKING, G.C. (1979b) Short-term toxicity of photomirex in the rat. Toxicol. appl. Pharmacol., 47: 105-114.

VILLENEUVE, D.C., VALLI, V.E., CHU, I., SECOURS, V., RITTER, L., & BECKING, G.C. (1979c) Ninety-day toxicity of photomirex in the male rat. Toxicology, 12: 235-250.

WALKER, E.M., Jr, GALE, G.R., ATKINS, L.M., & GADSDEN, R.H. (1977) Some effects of dieldrin and mirex on Ehrlich ascites tumor cells _in vivo_ and _in vitro_. _Arch. environ. Contam. Toxicol._, _5_: 333-341.

WARE, G.W. & GOOD, E.E. (1967) Effects of insecticides on reproduction in the laboratory mouse. II. Mirex, telodrin and DDT. _Toxicol. appl. Pharmacol._, _10_: 54-61.

WARREN, R.J., KIRKPATRICK, R.L., & YOUNG, R.W. (1978) Barbiturate-induced sleeping times, liver weights, and reproduction of cottontail rabbits after mirex ingestion. _Bull. environ. Contam. Toxicol._, _19_: 223-228.

WATERS, E.M. (1976) _Mirex: an overview and abstracted literature collection, 1947-1976_, Oak Ridge, Oak Ridge National Laboratory, Toxicology Information Response Center, 98 pp (ORNL/TIRC-76/4).

WIENER, M., PITTMAN, K.A., & STEIN, V. (1976) Mirex kinetics. _Drug Metab. Dispos._, _4_: 281-287.

WILKINSON, R.R, LAWLESS, E.W, MEINERS, A.F, FERGUSON, T.L, KELSO, G.L, & HOPKINS, S.C. (1978) State of the art report on pesticide disposal research. In: Kennedy, M.V., ed. _Disposal and decontamination of pesticides. A Symposium at the 174th Meeting of ACS, Chicago, August 29 - September 2, 1977_, pp. 73-79 (American Chemical Society Symposium Series 73).

WOJCIK, D.P., BANKS, W.A., WHEELER, W.B., JOUVENAZ, D.P., VAN MIDDELEM, C.H., & LOFGREN, C.S. (1975) Mirex residues in non-target organisms after application of experimental baits for fire ant control - South West Georgia 1971-1972. _Pestic. Monit. J._, _9_: 124-133.

WOLFE, J.L., ESHER, R.J., ROBINSON, K.M., & YARBROUGH, J.D. (1979) Lethal and reproductive effects of dietary mirex and DDT on old-field mice, _Peromyscus polionotus_. _Bull. environ. Contam. Toxicol._, _21_: 397-402.

WOODHAM, D.W., BOND, C.A., AHRENS, A.H., & MEDLEY, J.G. (1975) The cumulation and disappearance of mirex residues. III. In eggs and tissues of hens fed two concentrations of the insecticide in their diet. _Bull. environ. Contam. Toxicol._, _14_: 98-104.